T0212973

SpringerBriefs in Applied Sciences and Technology

Mathematical Methods

Series editor

Anna Marciniak-Czochra, Heidelberg, Germany

More information about this series at http://www.springer.com/series/11219

Ryszard Rudnicki · Marta Tyran-Kamińska

Piecewise Deterministic Processes in Biological Models

 Springer

Ryszard Rudnicki
Institute of Mathematics
Polish Academy of Sciences
Katowice
Poland

Marta Tyran-Kamińska
Institute of Mathematics
University of Silesia
Katowice
Poland

ISSN 2191-530X ISSN 2191-5318 (electronic)
SpringerBriefs in Applied Sciences and Technology
SpringerBriefs in Mathematical Methods
ISBN 978-3-319-61293-5 ISBN 978-3-319-61295-9 (eBook)
DOI 10.1007/978-3-319-61295-9

Library of Congress Control Number: 2017945225

Printed on acid-free paper

This Springer imprint is published by Springer Nature
The registered company is Springer International Publishing AG
The registered company address is: Gewerbestrasse 11, 6330 Cham, Switzerland

Preface

The aim of this book is to give a short mathematical introduction to piecewise deterministic Markov processes (PDMPs) and to present biological models where they appear. The book is divided into six chapters. In the first chapter we present some examples of biological phenomena such as gene activity and population growth leading to different type of PDMPs: continuous-time Markov chains, deterministic processes with jumps, dynamical systems with random switching and point processes. The second chapter contains some theoretical results concerning Markov processes and the construction of PDMPs. The next chapter is an introduction to the theory of semigroups of linear operators which provide the primary tools in the study of continuous-time Markov processes. In the fourth chapter we introduce stochastic semigroups, provide some theorems on their existence and find generators of semigroups related to PDMPs considered in the first chapter. The next chapter is devoted to the long-time behaviour (asymptotic stability and sweeping) of the stochastic semigroups induced by PDMPs. In the last chapter we apply the general results, especially concerning asymptotic behaviour, to biological models.

The book is dedicated to both mathematicians and biologists. The first group will find here new biological models which lead to interesting and often new mathematical questions. Biologists can observe how to include seemingly different biological processes into a unified mathematical theory and deduce from this theory interesting biological conclusions. We try to keep the required mathematical and biological background to a minimum so that the topics are accessible to students.

Acknowledgements

This research was partially supported by the National Science Centre (Poland) Grant No. 2014/13/B/ST1/00224. The authors are grateful to Michael C. Mackey and Katarzyna Pichór who read the manuscript and made useful suggestions for improvements.

Katowice, Poland   Ryszard Rudnicki
May 2017   Marta Tyran-Kamińska

Contents

Chapter 1
Biological Models

1.1 Introduction

Applications of probability theory to study biological and medical questions have a long history. Likely, probabilistic methods were used for the first time in the biological and medical sciences in 1766 by Daniel Bernoulli to demonstrate the efficacy of vaccination against smallpox [14]. From that time on, many problems in biology were described using probabilistic tools. Examples of applications of birth-death and branching processes in population dynamics and Bayesian methods in Mendelian genetics are well known [37]. Modern biology uses more advanced methods of probability theory to describe and study environmental and molecular processes. A special role in such applications is played by piecewise deterministic Markov processes (PDMPs).

According to a non-rigorous definition by Davis [28], the class of PDMPs is a general family of stochastic models covering virtually all non-diffusion applications. A formal definition of PDMP will be given at the end of this chapter. For now it is enough to know that PDMP is a continuous-time Markov process $\xi(t)$ with values in some metric space X and there is an increasing sequence of random times (t_n), called jump times, such that sample paths (trajectories) of $\xi(t)$ are defined in a deterministic way in each interval (t_n, t_{n+1}). We consider two types of behaviour of the process at jump times: the process can jump to a new point or can change the dynamics which defines its trajectories. PDMPs is a large family of different stochastic processes which includes discrete-time Markov processes, continuous-time Markov chains, deterministic processes with jumps, dynamical systems with random switching, stochastic billiards and some point processes.

In this chapter, we present simple biological models to illustrate possible applications of such processes. Some of these models were presented in the survey paper [101], which can serve as a short introduction to this monograph. Although discrete-time Markov processes play important role in applications we will not investigate them in this book because their theory differs from that of continuous-time PDMPs and their applications are sufficiently known [1].

© The Author(s) 2017   1
R. Rudnicki and M. Tyran-Kamińska, *Piecewise Deterministic Processes in Biological Models*, SpringerBriefs in Mathematical Methods,
DOI 10.1007/978-3-319-61295-9_1

1.2 Birth-Death Processes

We precede the definition of a birth-death process by a simple example concerning a cellular population. Consider a homogeneous population of cells and assume that during the time interval of length Δt a single cell can replicate with probability $b\Delta t + o(\Delta t)$ and it can die with probability $d\Delta t + o(\Delta t)$. We also assume that cells replicate and die independently of each other. Denote by $\xi(t)$ the number of cells at time t. Then for sufficiently small Δt the process $\xi(t)$ satisfies

$$\text{Prob}(\xi(t + \Delta t) = j | \xi(t) = i) = \begin{cases} bi\,\Delta t + o(\Delta t), & \text{if } j = i + 1, \\ di\,\Delta t + o(\Delta t), & \text{if } j = i - 1, \\ 1 - (b + d)i\,\Delta t + o(\Delta t), & \text{if } j = i, \\ o(\Delta t), & \text{if } |j - i| > 1. \end{cases} \tag{1.1}$$

It is assumed that $\lim_{\Delta t \to 0} \frac{o(\Delta t)}{\Delta t} = 0$. The process $\xi(t)$ is a continuous-time Markov chain with values in the space $\mathbb{N} = \{0, 1, 2, \ldots\}$. The process $\xi(t)$ is time homogeneous because the conditional probabilities (1.1) do not depend on t. The functions

$$p_{ji}(\Delta t) = \text{Prob}(\xi(t + \Delta t) = j | \xi(t) = i)$$

are called *transition probabilities* and the limit

$$q_{ji} = \lim_{\Delta t \to 0} \frac{p_{ji}(\Delta t)}{\Delta t}, \quad i \neq j,$$

is called the *transition rate* from i to j. The process $\xi(t)$ has transition rates $q_{i+1,i} = bi$, $q_{i-1,i} = di$, and $q_{ji} = 0$ otherwise. We also define

$$q_{ii} = -\sum_{j=0,\, j \neq i}^{\infty} q_{ji}. \tag{1.2}$$

The infinite matrix $Q = [q_{ji}]$ is called the *transition rate matrix* or the *infinitesimal generator matrix*. The matrix Q plays a crucial role in the theory of Markov processes. In particular, if $p_i(t) = \text{Prob}(\xi(t) = i)$ for $i \in \mathbb{N}$ and $t \geq 0$, then the column vector $p(t) = [p_0(t), p_1(t), \ldots]^T$ satisfies the following infinite dimensional system of differential equations

$$p'(t) = Qp(t).$$

The process $\xi(t)$ is one of many examples of birth-death processes. A general *birth-death process* is a continuous-time Markov chain defined by (1.1) if we replace bi with b_i and di with d_i, where (b_i) and (d_i) are sequences of non-negative numbers (see Fig. 1.1). The coefficients b_i and d_i are called *birth* and *death* rates for a

Fig. 1.1 The diagram of the transition between states in the birth-death process

population with i individuals. The problem of the existence of such a process is non-trivial and we shall consider it in detail in the next chapter. It should be noted that for given rates we have a family of different birth-death processes and their distributions depend on the initial distributions (i.e. the distribution of the random variables $\xi(0)$).

Two special types of birth-death processes are *pure birth* (if $d_i = 0$ for all i) and *pure death* (if $d_i = 0$ for all i) *processes*. A *Poisson process* $N(t)$ with intensity λ is a pure birth process with $b_i = \lambda$ for all i.

A similar construction based on the transition rate matrix can be applied to define general continuous-time Markov chains. Let $Q = [q_{ji}]$ be a transition rate matrix, i.e. a matrix which has non-negative entries outside the main diagonal and satisfies (1.2). The matrix Q is a transition rate matrix for a homogeneous continuous-time Markov chain $\xi(t)$ if the following condition is fulfilled

$$\mathrm{Prob}(\xi(t + \Delta t) = j | \xi(t) = i) = q_{ji}\Delta t + o(\Delta t).$$

Many applications of continuous-time Markov chains in population dynamics, epidemiology and genetics are given in Chap. VII of [1].

1.3 Grasshopper and Kangaroo Movement

Continuous-time Markov chains belong to a slightly larger class called pure jump-type Markov processes. A *pure jump-type Markov process* is a Markov process, which remains constant between jumps. For example, the process used in a simple description of the grasshopper and kangaroo movement [81] is an example of a pure jump-type Markov process, which is not a Markov chain. A grasshopper jumps at random times (t_n) from a point x to the point $x + \vartheta_n$. We assume that jump times are the same as for a Poisson process $N(t)$ with intensity $\lambda > 0$, i.e. $N(t_n) = n$, and that (ϑ_n) is a sequence of independent and identically distributed (i.i.d.) random vectors, which are also independent of $\{N(t)\colon t \geq 0\}$. Then the position $\xi(t)$ of the grasshopper is given by

$$\xi(t) = \xi(0) + \sum_{n=1}^{N(t)} \vartheta_n. \tag{1.3}$$

The process given by (1.3) is called a *compound Poisson process*.

A general *pure jump-type homogeneous Markov process* on a measurable space (X, Σ) can be defined in the following way. Let $\varphi: X \to [0, \infty)$ be a given measurable function and let $P(x, B)$ be a given *transition probability* on X, i.e. $P(x, \cdot)$ is a probability measure for each $x \in X$ and the function $x \mapsto P(x, B)$ is measurable for each $B \in \Sigma$. Let $t_0 = 0$ and let $\xi(0) = \xi_0$ be an X-valued random variable. For each $n \geq 1$ one can choose the *n*th *jump time* t_n as a positive random variable satisfying

$$\text{Prob}(t_n - t_{n-1} \leq t | \xi_{n-1} = x) = 1 - e^{-\varphi(x)t}, \quad t \geq 0,$$

and define

$$\xi(t) = \begin{cases} \xi_{n-1} & \text{for } t_{n-1} \leq t < t_n, \\ \xi_n & \text{for } t = t_n, \end{cases}$$

where the *n*th *post-jump position* ξ_n is an X-valued random variable such that

$$\text{Prob}(\xi_n \in B | \xi_{n-1} = x) = P(x, B).$$

1.4 Velocity Jump Process

Another type of cellular or organism dispersal considered in biological literature, e.g. in [81], is a *velocity jump process*. An individual is moving in the space $\mathbb{R}^d$ with a constant velocity and at jump times (t_n) it chooses a new velocity. As in Sect. 1.3, we assume that jump times are the same as for a Poisson process $N(t)$ with intensity λ. It means that $F(t) = 1 - e^{-\lambda t}$ is the distribution function of $t_n - t_{n-1}$. Let $x(t)$ be the position and $v(t)$ be the velocity of an individual at time t. We assume that for every $x, v \in \mathbb{R}^d$, there is a probability Borel measure $P(x, v, B)$ on $\mathbb{R}^d$ which describes the change of the velocity after a jump, i.e.

$$\text{Prob}(v(t_n) \in B \,|\, x(t_n^-) = x, \, v(t_n^-) = v) = P(x, v, B)$$

for every Borel subset B of $\mathbb{R}^d$, where $x(t_n^-)$ and $v(t_n^-)$ are the left-hand side limits of $x(t)$ and $v(t)$, respectively, at the point t_n. Then $\xi(t) = (x(t), v(t)), t \geq 0$, is a PDMP corresponding to our movement and between jumps the pair $(x(t), v(t))$ satisfies the following system of ordinary differential equations

$$\begin{cases} x'(t) = v(t), \\ v'(t) = 0. \end{cases} \tag{1.4}$$

There are a number of interesting examples of velocity jump processes. The simplest one is symmetric movement on the real line $\mathbb{R}$. In this case we assume that an individual is moving with constant speed, say one, and at a jump time it changes the direction of movement to the opposite one. A PDMP corresponding to the symmetric movement has values in the space $\mathbb{R} \times \{-1, 1\}$ and $P(x, v, \{-v\}) = 1$ for $v = -1, 1$. More advanced examples of velocity jump processes and their comparison with dispersal of cells, insects and mammals are given in [81, 109]. We can consider velocity jump processes defined in a bounded domain G. Examples of such processes are stochastic billiards [36]. Stochastic billiards do not change velocity in the interior of G but when an individual or a point strikes the boundary, a new direction is chosen randomly from the directions that point back into the interior of G and the motion continues. Stochastic billiards can describe not only the movement of individuals in a bounded domain but also various different biological processes. In Sect. 1.7, we present Lebowitz-Rubinow cell cycle model which can be identified with a one-dimensional stochastic billiard. Some examples of PDMPs with jumps on the boundary that appear in the theory of the gene regulatory systems are presented in Sect. 1.12.

There have been a number of papers devoted to velocity jump processes, their diffusion approximations and applications to aggregation and chemotaxis phenomena. The interested reader is referred to [50, 75, 82] and the references therein.

1.5 Size of Cells in a Single Line

We consider a sequence of consecutive descendants of a single cell. Let t_n be a time when a cell from the n-generation splits and let $\xi(t), t \in [t_{n-1}, t_n)$, be the size (mass, volume) of the cell from the n-generation at time t. We assume that $g(x)$ is the *growth rate* of a cell with size x, i.e. the process $\xi(t)$ satisfies the differential equation

$$\xi'(t) = g(\xi(t)) \quad \text{for } t \in (t_{n-1}, t_n), \ n \geq 1. \tag{1.5}$$

Denote by $\varphi(x)$ the *division rate* of a cell with size x, i.e. a cell with size x replicates during a small time interval of length Δt with probability $\varphi(x)\Delta t + o(\Delta t)$. Finally, we assume that a daughter cell has a half size of the mother cell. It means that

$$\xi(t_n) = \tfrac{1}{2}\xi(t_n^-), \tag{1.6}$$

where $\xi(t_n^-)$ is the left-hand side limit of $\xi(t)$ at the point t_n. In this way, we define a homogeneous PDMP which is not a pure jump-type Markov process.

Given growth and division rate functions and the initial size x_0 of a cell it is not difficult to find the distribution of its life-span and the distribution of its size at the point of division. Let $\pi(t, x_0)$ be the size of a cell at age t if its initial size is x_0, i.e. $\pi(t, x_0) = x(t)$, where x is the solution of the equation $x' = g(x)$ with the initial condition $x(0) = x_0$. The *life-span* of a cell is a random variable T which depends on the initial size x_0 of a cell. Let $\Phi_{x_0}(t) = \text{Prob}(T > t)$ be the *survival function*, i.e. the probability that the life-span of a cell with initial size x_0 is greater than t. Then

$$\text{Prob}(t < T \le t + \Delta t \mid T > t) = \frac{\Phi_{x_0}(t) - \Phi_{x_0}(t + \Delta t)}{\Phi_{x_0}(t)} = \varphi(\pi(t, x_0))\Delta t + o(\Delta t).$$

From this equation, we obtain

$$\Phi'_{x_0}(t) = -\Phi_{x_0}(t)\varphi(\pi(t, x_0))$$

and after simple calculations we get

$$\Phi_{x_0}(t) = \exp\left\{ - \int_0^t \varphi(\pi(s, x_0))\, ds \right\} \tag{1.7}$$

and $F(t) = 1 - \Phi_{x_0}(t)$ is the distribution function of the life-span. For $y \ge x_0$, we define $t(x_0, y)$ to be the time t such that $\pi(t, x_0) = y$. Since

$$\frac{\partial t}{\partial y} \cdot g(\pi(t, x_0)) = 1,$$

we see that

$$\frac{\partial t}{\partial y} = \frac{1}{g(y)}$$

and

$$\frac{\partial}{\partial y}\left(\int_0^{t(x_0, y)} \varphi(\pi(s, x_0))\, ds \right) = \frac{\varphi(y)}{g(y)}.$$

Let ϑ be the size of the cell at the moment of division. Then

$$\text{Prob}(\vartheta > y) = \text{Prob}(\pi(t, x_0) > y) = \exp\left\{ - \int_0^{t(x_0, y)} \varphi(\pi(s, x_0))\, ds \right\}$$

$$= \exp\left(- \int_{x_0}^y \frac{\varphi(r)}{g(r)}\, dr \right) = \exp(Q(x_0) - Q(y)), \tag{1.8}$$

where $Q(x) = \int_0^x \frac{\varphi(r)}{g(r)}\, dr$. Let ζ be a random variable with exponential distribution i.e. $\text{Prob}(\zeta > x) = e^{-x}$ for $x \ge 0$. Then

$$\text{Prob}(\vartheta > y) = \exp(Q(x_0) - Q(y)) = \text{Prob}\left(\zeta > Q(y) - Q(x_0) \right)$$

$$= \text{Prob}\left(Q^{-1}(Q(x_0) + \zeta) > y \right),$$

which means that the random variables ϑ and $Q^{-1}(Q(x_0) + \zeta)$ have the same distribution. From this it follows that

$$\xi(t_n) \stackrel{d}{=} \tfrac{1}{2}Q^{-1}(Q(\xi(t_{n-1})) + \zeta), \tag{1.9}$$

Fig. 1.2 A flow with jumps

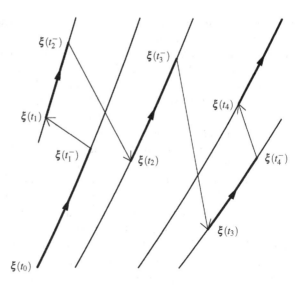

where ζ is a random variable independent of $\xi(t_{n-1})$ and with exponential distribution.

One can consider a more general model when a cell is characterized by a multidimensional parameter $x \in x \subset \mathbb{R}^d$ which changes according to a system of differential equations

$$x' = g(x),$$

$\varphi(x)$ is the division rate of a cell with parameter x, and $P(x, B)$ is a transition probability function on X which describes the change of parameters from a mother to a daughter cell, i.e.

$$\mathrm{Prob}(\xi(t_n) \in B | \xi(t_n^-) = x) = P(x, B).$$

The distribution function of $t_n - t_{n-1}$ is given by $F(t) = 1 - \Phi_{x_0}(t)$, where $\Phi_{x_0}(t)$ is as in (1.7) with $x_0 := \xi(t_{n-1})$. PDMPs of this type are called *flows with jumps* (see Fig. 1.2).

1.6 Two-Phase Cell Cycle Model

The cell cycle is a series of events that take place in a cell leading to its replication. Usually the cell cycle is divided into four phases. The first one is the growth phase G_1 with synthesis of various enzymes. The duration of the phase G_1 is highly variable even for cells from one species. The DNA synthesis takes place in the second phase S. In the third phase G_2 significant protein synthesis occurs, which is required during the process of mitosis. The last phase M consists of nuclear division and cytoplasmic division. From a mathematical point of view, we can simplify the model by considering only two phases: $A = G_1$ with a random duration t_A and B which consists of the phases S, G_2 and M. The duration t_B of the phase B is almost constant. There are several models of the cell cycle. Let us mention the models by Lasota and Mackey [61], Tyson and Hannsgen [112], and the generalization of these model given by Tyrcha [115]. Here, we describe a continuous-time version of the Tyrcha model and we show that it can be treated as a PDMP.

The crucial role in the model is played by a parameter x which describes the state of a cell in the cell cycle. It is not clear what x exactly should be. We simply assume that x is the cell size. We start with a model similar to that from Sect. 1.5 with only one difference. Here $\varphi(x)$ is the rate of entering the phase B, i.e. a cell with size x enters the phase B during a small time interval of length Δt with probability $\varphi(x)\Delta t + o(\Delta t)$. It is clear that the process $\xi(t)$ is piecewise deterministic but it is non-Markovian, because its future $\xi(t)$, $t \geq t_0$, depends not only on the random variable $\xi(t_0)$ but also depends in which phase it is at the time t_0.

Now we extend the process $\xi(t)$, $t \geq 0$, to obtain a homogeneous PDMP. A new process $\widetilde{\xi}(t)$, $t \geq 0$, is defined on the phase space $[0, \infty) \times [0, t_B] \times \{1, 2\}$ in the following way. First, we add additional jump points $s_n = t_n - t_B$, $n \geq 1$. At time s_n a cell from the n-generation enters the phase B (see Fig. 1.3). Let

$$\widetilde{\xi}(t) = (\widetilde{\xi}_1(t), \widetilde{\xi}_2(t), \widetilde{\xi}_3(t)) = (\xi(t), y(t), i),$$

where $i = 1$ if at time t a cell is in the phase A and $i = 2$ if it is in the phase B, $y(t) = 0$ if the cell is in the phase A and $y(t) = t - s_n$ if the cell from the n-generation is in the phase B. Between jump points the coordinates of the process $\widetilde{\xi}(t)$ satisfy the following system of ordinary differential equations

Fig. 1.3 The schematic diagram of the transition in the two-phase cell cycle model

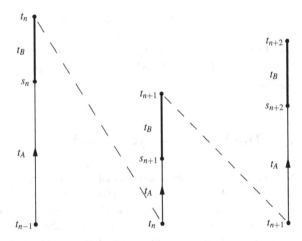

$$\begin{cases} \widetilde{\xi}_1'(t) = g(\widetilde{\xi}_1(t)), \\ \widetilde{\xi}_2'(t) = \begin{cases} 0, & \text{if } \widetilde{\xi}_3(t) = 1, \\ 1, & \text{if } \widetilde{\xi}_3(t) = 2, \end{cases} \\ \widetilde{\xi}_3'(t) = 0. \end{cases} \tag{1.10}$$

The transitions at the jump points are given by

$$\widetilde{\xi}_1(s_n) = \widetilde{\xi}_1(s_n^-), \quad \widetilde{\xi}_2(s_n) = \widetilde{\xi}_2(s_n^-) = 0, \quad \widetilde{\xi}_3(s_n) = 2,$$

and

$$\widetilde{\xi}_1(t_n) = \tfrac{1}{2}\widetilde{\xi}_1(t_n^-), \quad \widetilde{\xi}_2(t_n) = 0, \quad \widetilde{\xi}_3(t_n) = 1.$$

The distribution function of $s_n - t_{n-1}$ is given by

$$F(t) = 1 - \exp\left\{ -\int_0^t \varphi(\pi(s, x_0))\, ds \right\},$$

where $x_0 = \widetilde{\xi}_1(t_{n-1})$ and $\pi(t, x_0) = x(t)$ is the solution of equation $x' = g(x)$ with initial condition $x(0) = x_0$, while that of $t_n - s_n$ by $F(t) = 0$ for $t \le t_B$ and $F(t) = 1$ for $t > t_B$.

The life-span $t_n - t_{n-1}$ of a cell with initial size x_0 has the distribution

$$F(t) = \begin{cases} 0, & \text{if } t < t_B, \\ 1 - \exp\left\{ -\int_0^{t-t_B} \varphi(\pi(s, x_0))\, ds \right\}, & \text{if } t \ge t_B, \end{cases} \tag{1.11}$$

and the relation between the distributions of the random variables $\xi(t_n)$ and $\xi(t_{n-1})$:

$$\xi(t_n) \overset{d}{=} \tfrac{1}{2}\pi\left(t_B, \left(Q^{-1}(Q(\xi(t_{n-1})) + \zeta)\right)\right), \tag{1.12}$$

where ζ is a random variable independent of $\xi(t_{n-1})$ with exponential distribution.

1.7 Stochastic Billiard as a Cell Cycle Model

One of the oldest models of cell cycle introduced by Rubinow [94] is based on the concept of *maturity* and *maturation velocity*. Maturity, also called *physiological age*, is a real variable x from the interval $[0, 1]$ which describes the position of a cell in the cell cycle. A new born cell has maturity 0 and a cell splits at maturity 1. In Rubinow's model x grows according to the equation $x' = v$, where the maturation velocity v can depend on x and also on other factors such as time, the size of the population, temperature, light, environmental nutrients, pH, etc. If we neglect environmental factors, resource limitations, and stochastic variation, and we assume that v depends only on x, then all cells will have identical cell cycle in this model, in particular they have the same cell cycle length l.

However, experimental observations concerning cell populations, cultured under identical conditions for each member, revealed high variability of l in the population. It means that the population is heterogeneous with respect to cell maturation velocities, and therefore, mathematical models of the cell cycle should take into account maturation velocities. A model of this type was proposed by Lebowitz and Rubinow [64]. In their model the cell cycle is determined by its length l, which is fixed at the birth of the cell. The relation between cycle lengths of mother's and daughter's cells is given by a transition probability. If the maturation velocity of each individual cell is constant, we can use v instead of l in the description of the model because $lv = 1$. Such a model is a special case of the Rotenberg model [93] and we briefly present it here using PDMP formalism. Each cell matures with individual velocity v, which is a positive constant. We assume that the relation between the maturation velocities of mother's and daughter's cells is given by a transition probability $P(v, dv')$. As in Sect. 1.5, we consider a sequence of consecutive descendants of a single cell. Let t_n be a time when a cell from the n-generation splits and let $\xi(t) = (x(t), v(t))$, $t \in [t_{n-1}, t_n)$, be the maturity and the maturation velocity of the cell from the n-generation at time t. It is clear that $\xi(t), t \geq 0$, is a PDMP and between jumps the pair $(x(t), v(t))$ satisfies system (1.4). We assume that the process $\xi(t)$, $t \geq 0$, has càdlàg sample paths, i.e. they are right-continuous with left limits. We have $x(t_n) = 0$ and $\text{Prob}(v(t_n) \in B \,|v(t_n^-) = v) = P(v, B)$ for each $n \in \mathbb{N}$ and each Borel subset B of $(0, \infty)$. Since the $v(t)$ is constant in the interval (t_{n-1}, t_n), we have $t_n - t_{n-1} = 1/v(t_{n-1})$.

Properties of the density of the population with respect to (x, v) in the above model were studied in [19]. But this model can be also identified with a one-dimensional stochastic billiard. Namely, consider a particle moving in the interval $[0, 1]$ with a constant velocity. We assume that when the particle hits the boundary points 0 and 1, it changes its velocity according to the probability measures $P_0(-v, B) = P(v, B)$ and $P_1(v, -B) = P(v, B)$, respectively, where $v > 0$ and B is a Borel subset of $(0, \infty)$. Observe that the PDMP defined in the Lebowitz–Rubinow model is given by $\xi(t) = (x(t), |v(t)|)$, where $x(t)$ and $v(t)$ represent position and velocity of the moving particle at time t. Asymptotic properties of the general one-dimensional stochastic billiard were studied in [78].

It is not difficult to build a model which connects the classical Rubinow model with Lebowitz–Rubinow's one. We still assume that the cell maturity changes from 0 to 1 with velocity v and v depends on x and some specific genetic factor y (say phenotypic trait) inherited from the mother cell which does not change during the life of the cell. The trait y' of the daughter cell is drawn from a distribution $P(y, dy')$, where y is the mother cell trait. Similarly to the previous model, let t_n be a time when a cell from the n-generation splits and let $\xi(t) = (x(t), y(t))$, $t \in [t_{n-1}, t_n)$, be the maturity and the trait of the cell from the n-generation at time t. Then between jumps the pair $(x(t), y(t))$ satisfies the following system:

$$\begin{cases} x'(t) = v(x(t), y(t)), \\ y'(t) = 0. \end{cases} \tag{1.13}$$

We have also $x(t_n) = 0$ and $\text{Prob}(y(t_n) \in B \,|\, y(t_n^-) = y) = P(y, B)$. Then the length of the interval $T = t_n - t_{n-1}$ depends on the trait y of the cell and T is a positive number such that $x(T) = 1$, where $x(t)$ is the solution of the initial problem $x'(t) = v(x(t), y), x(0) = 0$.

As we mentioned above, the Lębowitz–Rubinow model is a special case of the Rotenberg model [93]. Now we briefly present this model using PDMP formalism. In the Rotenberg model, the maturation velocity is random and can change also during the cell cycle. A new born cell inherits the initial maturation velocity from its mother as in the Lebowitz–Rubinow model according to a transition probability $P(v, dv')$. During the cell cycle it can change its maturation velocity with intensity $q(x, v)$, i.e. a cell with parameters (x, v) can change the maturation velocity in a small time interval of length Δt with probability $q(x, v)\Delta t + o(\Delta t)$. Let x_0 be the maturity of a cell when it receives the maturation velocity v. From (1.7) it follows that the probability distribution function of the length of the time interval when the maturation velocity remains v is given by

$$1 - \exp\left\{ -\int_0^t q(x_0 + vs, v)\, ds \right\}.$$

We suppose that if (x, v) is the state of the cell at the moment of division, then a new maturation velocity is drawn from a distribution $P(x, v, dv')$. The process $\xi(t) = (x(t), v(t))$ describing consecutive descendants of a single cell is a PDMP which has jumps when cells split and random jumps during their cell cycles. Between jumps the pair $(x(t), v(t))$ satisfies system (1.4). If a jump is at the moment of division then it is given by the same formula as in the Lebowitz–Rubinow model. If a jump is during the cell cycle, then $x(t_n) = x(t_n^-)$ and

$$\text{Prob}(v(t_n) \in B \,|\, x(t_n^-) = x, \ v(t_n^-) = v) = P(x, v, B) \text{ for each Borel subset } B \text{ of } (0, \infty).$$

1.8 Stochastic Gene Expression I

Gene expression is a complex process which involves three processes: gene activation/inactivation, mRNA transcription/decay and protein translation/decay. Now we consider a very simple model when proteins production is regulated by a single gene and we omit the intermediate process of mRNA transcription (see Fig. 1.4). A gene can be in an active or an inactive state and it can be transformed into an active state or into an inactive state, with intensities q_0 and q_1, respectively. The rates q_0 and q_1 depend on the number of protein molecules $\xi(t)$. If the gene is active then proteins

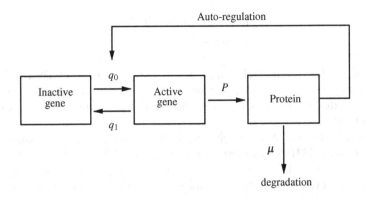

Fig. 1.4 The diagram of gene expression I

are produced with a constant speed P. In both states of the gene, protein molecules undergo the process of degradation with rate μ. It means that the process $\xi(t), t \geq 0$, satisfies the equation

$$\xi'(t) = PA(t) - \mu\xi(t), \qquad (1.14)$$

where $A(t) = 1$ if the gene is active and $A(t) = 0$ in the opposite case. Then the process $\widetilde{\xi}(t) = (\xi(t), A(t)), t \geq 0$, is a PDMP. Since the right-hand side of Eq. (1.14) is negative for $\xi(t) > \frac{P}{\mu}$ we can restrict values of $\xi(t)$ to the interval $\left[0, \frac{P}{\mu}\right]$ and the process $\widetilde{\xi}(t)$ is defined on the phase space $\left[0, \frac{P}{\mu}\right] \times \{0, 1\}$.

The process $\widetilde{\xi}(t)$ has jump points when the gene changes its activity. Formula (1.7) allow us to find the distribution of time between consecutive jumps. Observe that if x_0 is the number of protein molecules at a jump point, then after time t we have

$$\pi_t^0(x_0) = x_0 e^{-\mu t}, \quad \pi_t^1(x_0) = \frac{P}{\mu} + \left(x_0 - \frac{P}{\mu}\right)e^{-\mu t}$$

protein molecules, respectively, in an inactive and an active state. From (1.7), it follows that the probability distribution function of the length of an inactive state is given by

$$1 - \exp\left\{-\int_0^t q_0\left(x_0 e^{-\mu s}\right) ds\right\}$$

and of an active state by

$$1 - \exp\left\{-\int_0^t q_1\left(\frac{P}{\mu} + \left(x_0 - \frac{P}{\mu}\right)e^{-\mu s}\right) ds\right\}.$$

1.9 Stochastic Gene Expression II

A more advanced (and more realistic) model of gene expression was introduced by Lipniacki et al. [65] and studied in [17]. In this model, mRNA transcription and decay processes are also taken into consideration. The rates q_0 and q_1 depend on the number of mRNA molecules $\xi_1(t)$ and on the number of protein molecules $\xi_2(t)$. If the gene is active then mRNA transcript molecules are synthesized with a constant speed R. The protein translation proceeds with the rate $P\xi_1(t)$, where P is a constant (see Fig. 1.5). The mRNA and protein degradation rates are μ_R and μ_P, respectively. Now, instead of Eq. (1.14) we have the following system of differential equations

$$\begin{cases} \xi_1'(t) = RA(t) - \mu_R\xi_1(t), \\ \xi_2'(t) = P\xi_1(t) - \mu_P\xi_2(t), \end{cases} \tag{1.15}$$

where $A(t) = 1$ if the gene is active and $A(t) = 0$ in the opposite case. Then the process $\widetilde{\xi}(t) = (\xi_1(t), \xi_2(t), A(t))$, $t \geq 0$, is a homogeneous PDMP.

As in the previous model the process $\widetilde{\xi}(t)$ has jump points when the gene changes its activity and formula (1.7) allow us to find the distribution of the time between consecutive jumps. Let x_0 and y_0 be the initial number of mRNA and protein molecules. Let us denote by $\pi_t^i(x_0, y_0) = x(t)$ the solution of the system (1.15) at time t, where $i = 0$ and $i = 1$ correspond to an inactive and an active state, respectively (see Fig. 1.6). Then the probability distribution function of the length of an inactive (or active) state is given by

$$1 - \exp\left\{-\int_0^t q_i(\pi_s^i(x_0, y_0))\, ds\right\}.$$

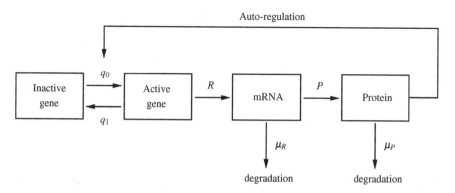

Auto-regulation

Fig. 1.5 The diagram of autoregulated gene expression II

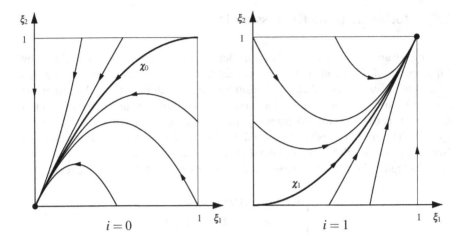

Fig. 1.6 The flows π^i for $P = R = \mu_P = \mu_R = 1$: *left $i = 0$, right $i = 1$*. The set between the curves χ_0 and χ_1 is a stochastic attractor of the related PDMP, i.e. almost all trajectories of the PDMP enter this set and do not leave it

In some cases the process of protein production has an additional primary transcript step [100]. After the activation of a gene, the DNA code is transformed into pre-mRNA form of transcript. Then, pre-mRNA molecules are converted to functional forms of mRNA, which is transferred into the cytoplasm, where during the translation phase mRNA is decoded into a protein. We consider three-dimensional model of gene expression with variables ξ_1, ξ_2, ξ_3 describing evolution of pre-mRNA, mRNA and protein levels. This model is described by the following system of differential equations

$$\begin{cases} \xi_1'(t) = RA(t) - (C + \mu_{pR})\xi_1(t), \\ \xi_2'(t) = C\xi_1(t) - \mu_R\xi_2(t), \\ \xi_3'(t) = P\xi_2(t) - \mu_P\xi_3(t). \end{cases} \tag{1.16}$$

Here R is the speed of synthesis of pre-mRNA molecules if the gene is active, C is the rate of converting pre-mRNA into active mRNA particles, μ_{pR} is the pre-mRNA degradation rate and the rest constants are the same as in the previous model.

Remark 1.1 Examples from Sects. 1.8 and 1.9 show that using PDMPs to model autoregulated genetic networks is very natural. New results concerning this subject can be also found in [71, 77, 122, 123].

Models described in the last two sections lead to the following general scheme called *dynamical systems with random switching* (see Fig. 1.7). We consider a family of deterministic processes indexed by a set $I \subset \mathbb{R}$. Each process is a solution of a system of differential equations in some subset X of the space $\mathbb{R}^d$:

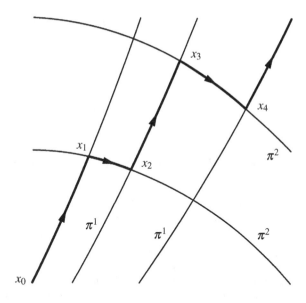

Fig. 1.7 A dynamical system with random switching: $x_1 = \pi^1_{T_1} x_0$, $x_2 = \pi^2_{T_2} x_1$, ...

$$x'(t) = b^i(x(t)), \qquad (1.17)$$

$i \in I$. Temporary, elements of I are called states. The set I should be a measurable set but it can be finite, countable or even uncountable. We assume that for each point $x \in X$ we have a transition probability on I denoted by $P_x(i, B)$. We assume that $q_i(x)$ is the intensity of leaving the state i, i.e. if at time t the system is in the state i and at the point $x \in \mathbb{R}^d$, then it leaves the state i in a small time interval of length Δt with probability

$$q_i(x)\Delta t + o(\Delta t).$$

We choose a point $x_0 \in \mathbb{R}^d$ and a state $i \in I$. We go along the trajectory, which is the solution of Eq. (1.17) with the initial condition $x(0) = x_0$. We stay in the state i for a random time T_i and the random variable T_i has the probability distribution function

$$1 - \exp\left\{-\int_0^t q_i(\pi^i_s(x_0))\, ds\right\}, \qquad (1.18)$$

where $\pi^i_t(x_0)$ is the solution of Eq. (1.17) with the initial condition $x(0) = x_0$. At time T_i we leave the state i and using transition probability $P_{x_1}(i, B)$, $x_1 = \pi^i_{T_i}(x_0)$, we choose the next state, say j. We repeat the procedure starting now from the point x_1 and the state j and continue it in subsequent steps. If $i(t)$ is the state of the system at time t then the process $\tilde{\xi}(t) = (x(t), i(t))$, $t \geq 0$, is a homogeneous PDMP. We can also treat the process $\tilde{\xi}(t)$ as a solution of a single system of equations:

$$\begin{cases} x'(t) = b^{i(t)}(x(t)), \\ i'(t) = 0, \end{cases} \tag{1.19}$$

with transition at the jump times given by

$$x(t_n) = x(t_n^-), \quad \text{Prob}(i(t_n) \in B) = P_{x(t_n^-)}(i(t_n^-), B).$$

1.10   Gene Regulatory Models with Bursting

The central dogma of molecular biology, suggested 60 years ago, proposes that the information flows from DNA to mRNA and then to proteins. A gene is expressed through the processes of transcription and translation. First, in the process of transcription, the enzyme RNA polymerase uses DNA as a template to produce an mRNA molecule which can be translated to make the protein encoded by the original gene. Recent advances to monitor the behaviour of single molecules allow experimentalists to quantify the transcription of mRNA from the gene as well as the translation of the mRNA into protein. The visualization revealed that in many cases mRNA and protein synthesis occur in quantal bursts in which the production is active for a relatively short and random period of time and during that time a random number of molecules is generated. There is the well-documented production of mRNA and/or protein in stochastic bursts in both prokaryotes and eukaryotes. For example, in [25] the authors observe bursts of expression of beta-galactosidase in E. coli, yeast and mouse cells.

An analysis of the data obtained from such experiments was used in [40] to predict the size and frequency of the bursts of proteins and to derive some of the first analytical expressions for protein distributions in steady states across a population of cells. Including mRNA dynamics in gene regulatory models introduces more complexity. We assume that the rates of degradation of mRNA and of protein are γ_1 and γ_2, respectively, and that k_1 is the rate of transcription, while k_2 is the rate of translation of mRNA. Let us first consider translational bursting. It was observed in [25, 121] that the amplitude of protein production through bursting translation of mRNA is geometric/exponentially distributed with mean number of protein molecules per burst equal to $b = k_2/\gamma_1$ and the frequency of bursts being equal to $a = k_1/\gamma_2$. If $\xi(t)$ denotes the number of protein molecules then $\xi(t)$ can be described in terms of a continuous-time Markov chain with state space equal to $\mathbb{N}$ (see [72]). Here, we follow the approach of [40] (see also [71, 72]) and we assume that $\xi(t)$ denotes the concentration of protein in a cell, which is the number of molecules divided by the volume of the cell, being a continuous variable. We consider the general situation when the degradation of proteins is interrupted at random times t_n when new molecules are being produced with frequency φ depending on the current level x of the concentration of proteins and according to a distribution with density $h(x, y)$. Thus between the jump times $\xi(t)$ is the solution of the equation

$$\xi'(t) = -\gamma_2 \xi(t)$$

and at jump time t_n we go from the point $\xi(t_n-)$ to the point $\xi(t_n-) + \vartheta_n$ where ϑ_n has a distribution with density h, possibly depending on x, so that

$$\text{Prob}(\vartheta_n \in B | \xi(t_n-) = x) = \int_B h(x, y) \, dy.$$

In particular, the model considered in [40] assumes that φ is a constant equal to a and that

$$h(x, y) = \frac{1}{b} e^{-y/b}, \quad y > 0,$$

in that case one can easily calculate a steady-state distribution, which is the gamma distribution with parameters a and b. To get analytical expressions for the distribution one can also take functions regulating the frequency of bursts to be Hill-type functions, see [40, 71, 72].

Transcriptional bursts have also been observed [43] with the number of molecules produced per burst being exponentially distributed. We let ξ_1 and ξ_2 denote the concentrations of mRNA and protein respectively. We assume that ξ_1 and ξ_2 are solutions of the equations

$$\xi_1'(t) = -\gamma_1 \xi_1(t), \quad \xi_2'(t) = k_2 \xi_1(t) - \gamma_2 \xi_2(t),$$

between jump times, and that mRNA are produced in bursts with jump rate φ depending on the current level ξ_2 of proteins, see [15]. Regulatory mechanisms and genetic switches can be modelled using this type of PDMPs [18, 69, 73, 80]. These are particular examples of the so-called flows with jumps as described at the end of Sect. 1.5.

1.11 Neural Activity

A neuron is an electrically excitable cell that processes and transmits information through electrical signals. The neuron's membrane potential V_m is the inside potential minus the outside potential. If a cell is in the resting state, then this potential, denoted by $V_{m,R}$, is about -70 mV. The *depolarization* is defined as

$$V = V_m - V_{m,R}.$$

A cell is said to be *excited* (or depolarized) if $V > 0$ and *inhibited* (or hyperpolarized) if $V < 0$. The Stein model [107, 108] describes how the depolarization $V(t)$ is changing in time. The cell is initially at rest so that $V(0) = 0$. Nerve cells may be excited or inhibited through neuron's synapses—junctions between nerve cells (or

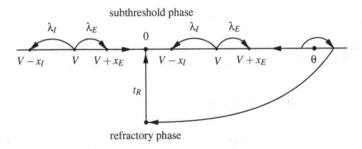

subthreshold phase

Fig. 1.8 Changes of the depolarization in the Stein model

between muscle and nerve cell) such that electrical activity in one cell may influence the electrical potential in the other. Synapses may be excitatory or inhibitory. We assume that there are two non-negative constants x_E and x_I such that if at time t an excitation occurs then $V(t^+) = V(t^-) + x_E$ and if an inhibition occurs then $V(t^+) = V(t^-) - x_I$. The jumps (excitations and inhibitions) may occur at random times according to two independent Poisson processes $N^E(t)$, $N^I(t)$, $t \geq 0$, with positive intensities λ_E and λ_I, respectively. Between jumps the depolarization $V(t)$ decays according to the equation $V'(t) = -\alpha V(t)$. When a sufficient (threshold) level $\theta > 0$ of excitation is reached, the neuron emits an action potential (fires). This will be followed by an absolute refractory period of duration t_R, during which $V \equiv 0$ and then the process starts again (see Fig. 1.8).

Now, we describe the neural activity as a PDMP. Since the refractory period has a constant duration we can use a model similar to that of Sect. 1.6 with two phases A and B, where A is the subthreshold phase and B is the refractory phase of duration t_R. We consider two types of jump points: when the neuron is excited or inhibited and the ends of refractory periods. Thus, we can have one or more jumps inside the phase A. Let $t_0, t_1, t_2, \ldots$ be the subsequent jump times. We denote by $\mathscr{F}$ the subset of jump points consisting of firing points. Let $\widetilde{\xi}(t) = (V(t), 1)$ if the neuron is in the phase A, and $\widetilde{\xi}(t) = (y, 2)$ if the neuron is in the phase B, where y is the time since the moment of firing. The process $\widetilde{\xi}(t) = (\widetilde{\xi}_1(t), \widetilde{\xi}_2(t))$, $t \geq 0$, is defined on the phase space $(-\infty, \theta) \times \{1\} \cup [0, t_R] \times \{2\}$ and between jumps it satisfies the following system of equations:

$$\widetilde{\xi}_1'(t) = \begin{cases} -\alpha\widetilde{\xi}_1(t), & \text{if } \widetilde{\xi}_2(t) = 1, \\ 1, & \text{if } \widetilde{\xi}_2(t) = 2. \end{cases}$$

If the neuron is in the phase A, i.e. $\widetilde{\xi}_2(t) = 1$, the depolarization can jump with intensity $\lambda = \lambda_E + \lambda_I$. It means that $F(t) = 1 - e^{-\lambda t}$ is the distribution function of $t_n - t_{n-1}$ if $t_{n-1} \notin \mathscr{F}$, while $F(t) = 0$ for $t < t_R$ and $F(t) = 1$ for $t \geq t_R$ if $t_{n-1} \in \mathscr{F}$. The transition at a jump point depends on the state of the neuron (its phase and the value of its depolarization). If $\widetilde{\xi}(t_n^-) = (t_R, 2)$ then $\widetilde{\xi}(t_n) = (0, 1)$ with probability one; if $\widetilde{\xi}(t_n^-) = (V, 1)$ and $V < \theta - x_E$ then $\widetilde{\xi}(t_n) = (V + x_E, 1)$ with probability

λ_E/λ and $\widetilde{\xi}(t_n) = (V - x_I, 1)$ with probability λ_I/λ; while if $\widetilde{\xi}(t_n^-) = (V, 1)$ and $V \geq \theta - x_E$ then $\widetilde{\xi}(t_n) = (0, 2)$ with probability λ_E/λ and $\widetilde{\xi}(t_n) = (V - x_I, 1)$ with probability λ_I/λ.

Remark 1.2 In [26] a simpler version of Stein's model is considered without the refractory period. A PDMP $V(t)$, $t \geq 0$, corresponding to this model is defined on the phase space $(-\infty, \theta)$, between jumps it satisfies the equation $V'(t) = -\alpha V(t)$, and $F(t) = 1 - e^{-\lambda t}$ is the distribution function of $t_n - t_{n-1}$. The transition at a jump point are given by two formulae:
if $V(t_n^-) < \theta - x_E$ then $V(t_n) = V(t_n^-) + x_E$ with probability λ_E/λ and $V(t_n) = V(t_n^-) - x_I$ with probability λ_I/λ;
and if $V(t_n^-) \geq \theta - x_E$ then $V(t_n) = 0$ with probability λ_E/λ and $V(t_n) = V(t_n^-) - x_I$ with probability λ_I/λ.

Remark 1.3 There are several different models of the neural activity. The interested reader is referred to the monographs [7, 110] and papers [23, 24, 39, 42, 111].

1.12 Processes with Extra Jumps on a Subspace

Many interesting applications of PDMPs in biological sciences can be found in books and papers devoted to stochastic hybrid systems. Such systems have appeared as stochastic versions of deterministic hybrid systems—systems described by differential equations and jumps. Stochastic hybrid systems are very close to PDMPs but they also contain diffusion processes. One can find a definition of stochastic hybrid systems and many examples of their applications in the books [3, 16, 27, 119] and the papers [51, 106].

In [52, 59], which are devoted to stochastic hybrid systems, we can find models of the production of subtilin by the bacterium *Bachillus subtilis*. Although this process plays an important role in biochemical technology, we do not recall these models in detail but only draw the reader's attention to an interesting mathematical aspect of these models. In order to survive, bacteria produce an antibiotic called subtilin to eliminate competing microbial species in the same ecosystem. The model considered in [52] consists of five differential equations describing: the size of *B. subtilis* population x_1, the total amount of nutrient available in the environment x_2, and the concentration of three types of proteins taking part in the process x_3, x_4, x_5. Two of these equations are without switches, two have switches depending on the states of two genes and one of them have a switch which depends on the level of the total amount of nutrient, namely,

$$x_3'(t) = \begin{cases} -\mu x_3(t), & \text{if } x_2(t) \geq \theta, \\ P_3 - \mu x_3(t), & \text{if } x_2(t) < \theta, \end{cases}$$

where θ is a positive constant. A PDMP corresponding to this model is defined on the space $\mathbb{R}_+^5$ and has extra jumps on the subspace $S = \{x \in \mathbb{R}_+^5 : x_2 = \theta\}$.

Processes with jumps at some surfaces appear in a natural way in other models. For example, if we consider the movement of an individual in a bounded domain G we can use a velocity jump process to describe this movement in the interior of this domain (see Sect. 1.4). But if the process reaches the boundary ∂G of this domain then we can use the theory of classical and stochastic billiards [36] to describe the changes of the velocity.

We now give two simple examples of processes with jumps on a subspace which allow us to reveal some problems with definition of PDMPs. The first one is a modification of the model of gene expression from Sect. 1.8. Again, we denote the number of protein molecules by $\xi(t)$ and assume that protein molecules degrade with rate μ and if the gene is active then proteins are produced with rate P, thus $\xi(t)$ satisfies Eq. (1.14). The gene can be inactivated with intensity $q_1(\xi(t))$ and activated if the number of protein molecules $\xi(t)$ reaches the minimal level θ, where $0 \le \theta < P/\mu$. The process $\widetilde{\xi}(t) = (\xi(t), A(t))$, $t \ge 0$, is a PDMP defined on the phase space $\left[\theta, \frac{P}{\mu}\right] \times \{0, 1\}$ and has extra jumps at the boundary point $(\theta, 0)$. Now, the intensity of activation cannot be a function, but, if we formally assume that this intensity is an infinite measure concentrated at the point θ, then formula (1.7) allows us to find the distribution of the time between consecutive jumps. If x_0 is the number of protein molecules at the last jump from active to inactive state, then $\pi_t^0(x_0) = x_0 e^{-\mu t}$ and $\pi_t^0(x_0) = \theta$ for $t = T := \frac{1}{\mu} \log(x_0/\theta)$. It means that the probability distribution function of the length of an inactive state is given by

$$F(t) = \begin{cases} 0, & \text{if } t < T, \\ 1, & \text{if } t \ge T. \end{cases} \qquad (1.20)$$

A similar modification can be made in the model from Sect. 1.9, i.e. we assume that the gene is activated if the number of protein molecules $\xi_2(t)$ reaches the level θ, where $0 < \theta \le PR/\mu_R\mu_P$. It can happen that $\xi_1(\bar{t}) < \mu_p\theta$ at the moment of gene activation $\bar{t}$, and then $\xi_2'(\bar{t}) < 0$. It implies that the number of protein molecules $\xi_2(t)$ decreases in some interval $(\bar{t}, \bar{t} + \varepsilon)$, thus it goes below the level θ. It is clear that the gene should remain active until the number of protein molecules $\xi_2(t)$ crosses at least the level θ, therefore, we assume that the intensity of inactivation $q_1(x_1, x_2)$ equals zero if $x_2 \le \theta$. The corresponding PDMP $\widetilde{\xi}(t) = (\xi_1(t), \xi_2(t), A(t))$, $t \ge 0$ is defined on the phase space $X = \left[0, \frac{R}{\mu_R}\right] \times \left[0, \frac{PR}{\mu_P\mu_R}\right] \times \{0, 1\}$ and has extra jumps at the line segment $S = \left[0, \frac{R}{\mu_R}\right] \times \{\theta\} \times \{0, 1\}$. Here, an extra jump can be inside the phase space. The probability distribution function of the length of an inactive state is given by Eq. (1.20) and the intensity of activation is a measure such that $\mu(X \setminus S) = 0$ and $\mu\{(x, \theta, 0)\} = \infty$ for $x \in \left[0, \frac{R}{\mu_R}\right]$.

An interesting PDMP with jumps on some subset is used in [20, 21] to describe the motor-driven transport of vesicular cargo to synaptic targets located on the axon or dendrites of a neuron. In this model, a particle is transported along the interval $[0, L]$ to some target x_0. The particle can be in one of three states $i \in I = \{-1, 0, 1\}$

and at a given state i it moves with speed i. Transitions between the three states are governed by a time homogeneous Markov chain $\vartheta(t)$ defined on the set I with the transition rate from i to j given by q_{ji}. The particle starts from the point $x = 0$ and moves to the right. We assume that 0 is a reflecting point and L is an absorbing point. Finally, we assume that if the particle is in some neighbourhood U of the target x_0 and is in the stationary state $i = 0$, then the particle can be absorbed by the target at a rate κ. If the particle is absorbed by the target or by the point L then the process repeats. The movement of the particle between jumps is described by a velocity jump process $\xi(t) = (x(t), i(t))$, $t \geq 0$, from Sect. 1.4 defined on the phase space $[0, L] \times I$, where the pair $(x(t), i(t))$ satisfies the following system of ordinary differential equations

$$\begin{cases} x'(t) = i(t), \\ i'(t) = 0. \end{cases} \tag{1.21}$$

If $x \in (0, L)$ then the process $\xi(t)$ can jump from (x, i) to (x, j) with intensity q_{ji} for $i, j \in I$, and if $x \in U$ the process $\xi(t)$ can also jump from $(x, 0)$ to $(0, 1)$ with intensity κ. The process $\xi(t)$ has two extra jumps from $(0, -1)$ and from $(L, 1)$ to $(0, 1)$ with probability 1.

There are a lot of different stochastic models of intracellular transport. The interested reader is referred to the review paper [22].

Remark 1.4 We can consider PDMPs with jump intensities given by measures defined on the trajectories of the processes. Assume that a process $\xi(t)$ is defined on the space $X \subset \mathbb{R}^d$ and between jumps satisfies the equation $\xi'(t) = g(\xi(t))$. Let $\pi(t, x_0) = x(t)$ be the solution of this equation with the initial condition $x(0) = x_0$ defined on the maximal time interval (a, b), $-\infty \leq a < b \leq \infty$. Let $L = \{\pi(t, x_0): t \in (a, b)\}$ be the trajectory of x_0. We consider only the case when the trajectory is not periodic. For $x = \pi(t_1, x_0)$ and $y = \pi(t_2, x_0)$ with $t_1 < t_2$, we denote by $L(x, y)$ the arc $\{\pi(t, x_0): t \in (t_1, t_2]\}$. The jump intensity on the trajectory L can be a non-negative Borel measure μ on L. If $\xi(0) = x$, $x \in L$, then the probability distribution function of the time till the next jump is given by

$$F(t) = 1 - \Phi_x(t) = 1 - e^{-\mu(L(x, \pi(t, x)))}.$$

In particular, if $\mu(x) = c > 0$ and $\xi(t^-) = x$ then the process has a jump at time t with probability $1 - e^{-c}$. In this case we should assume that the probability of jump at the point $y = \xi(t)$ is zero.

1.13 Size-Structured Population Model

We return back to the model of size distribution from Sect. 1.5 but instead of a single cell we consider the size distribution of all cells in the population. As in Sect. 1.5 the size $x(t)$ of a cell grows according to the equation

$$x'(t) = g(x(t)).$$

A single cell with size x replicates with rate $b(x)$ and dies with the rate $d(x)$. A daughter cell has a half size of the mother cell. Let assume that at time t we have k cells and denote their sizes by $x_1(t), x_1(t), \ldots, x_k(t)$. We can assume that a state of the population at time t is the set

$$\{x_1(t), \ldots, x_k(t)\}$$

and the evolution of the population is a stochastic process $\xi(t) = \{x_1(t), \ldots, x_k(t)\}$. Since the values of this process are sets of points, the process $\xi(t)$ is called a *point process*. Although such approach is a natural one, it has one important disadvantage. We are not able to describe properly the situation when two cells have the same size. One of the solution of this problem is to consider $\xi(t)$ as a process whose values are multisets. We recall that a *multiset* (or a *bag*) is a generalization of the notion of set in which members are allowed to appear more than once. Another solution of this problem is to consider $\xi(t)$ as a process with values in the space of measures given by

$$\xi(t) = \delta_{x_1(t)} + \cdots + \delta_{x_k(t)},$$

where δ_a denotes the *Dirac measure* at the point a, i.e. δ_a is the probability measure concentrated at the point a. Also this approach has some disadvantages, for example it is rather difficult to consider differential equations on measures. One solution of this problem is to consider a state of the system as a k-tuple $(x_1(t), \ldots, x_k(t))$. But some cells can die or split into two cells which changes the length of the tuple. To omit this problem we introduce an extra "dead state" $*$ and describe the state of the population at time t as an infinite sequence of elements from the space $\mathbb{R}_*^+ = [0, \infty) \cup \{*\}$ which has on some positions the numbers $x_1(t), \ldots, x_k(t)$ and on other positions has $*$. In order to have uniqueness of states, we introduce an equivalence relation $\sim$ in the space X of all $\mathbb{R}_*^+$-valued sequences x such that $x_i = *$ for all but finitely many i. Two sequences $x \in X$ and $y \in X$ are equivalent with respect to $\sim$ if y can be obtained as a permutation of x, i.e. $x \sim y$ if and only if there is a bijective function $\sigma : \mathbb{N} \to \mathbb{N}$ such that $y = (x_{\sigma(1)}, x_{\sigma(2)}, \ldots)$. The phase space $\widetilde{X}$ in our model is the space of all equivalence classes with respect to $\sim$, i.e. $\widetilde{X} = X/\sim$.

We now can describe the evolution of the population as a stochastic process $\xi(t) = [(x_1(t), x_2(t), \ldots)]$ with values in the space $\widetilde{X}$. The process $\xi(t)$ has jump points when one of the cells dies or replicates. We define $g(*) = b(*) = b(*) = 0$ and assume that $x(t) = *$ is the solution of the equation $x'(t) = 0$ with the initial condition $x(0) = *$. Between jumps the process $\xi(t)$ satisfies the equation

$$[(x_1'(t) - g(x_1(t)), x_2'(t) - g(x_2(t)), \ldots)] = [(0, 0, \ldots)]. \tag{1.22}$$

For $t \geq 0$ and $x^0 \in \mathbb{R}_*^+$ we denote by $\pi(t, x^0)$ the solution $x(t)$ of the equation $x'(t) = g(x(t))$ with the initial condition $x(0) = x^0$. Let $\mathbf{x}^0 = [(x_1^0, x_2^0, \ldots)] \in \widetilde{X}$ and define

$$\tilde{\pi}(t, \mathbf{x}^0) = [(\pi(t, x_1^0), \pi(t, x_2^0), \ldots)].$$

The jump rate function $\varphi(\mathbf{x})$ at the state $\mathbf{x} = [(x_1, x_2, \ldots)]$ is the sum of rates of death and division of all cells:

$$\varphi(\mathbf{x}) = \sum_{i=1}^{\infty} (b(x_i) + d(x_i)). \tag{1.23}$$

If $\mathbf{x}^0 \in \widetilde{X}$ is the initial state of the population at a jump time t_n, then the probability distribution function of $t_{n+1} - t_n$ is given by

$$1 - \exp \left\{ - \int_0^t \varphi(\tilde{\pi}(s, \mathbf{x}^0)\, ds \right\}. \tag{1.24}$$

At time t_n one of the cells dies or replicates. If a cell dies we change the sequence by removing its size from the sequence and we have

$$\mathrm{Prob}\left(\xi(t_n) = [(x_1(t_n^-), \ldots, x_{i-1}(t_n^-), x_{i+1}(t_n^-), x_{i+2}(t_n^-), \ldots)]\right) = \frac{d_i(x_i(t_n^-))}{\varphi(\xi(t_n^-))}$$

for $i \in \mathbb{N}$. If a cell replicates, we remove its size from the sequence and add two new elements in the sequence with the sizes of the daughter cells and we have

$$\mathrm{Prob}\left(\xi(t_n) = [(x_1(t_n^-), \ldots, x_{i-1}(t_n^-), \tfrac{1}{2}x_i(t_n^-), \tfrac{1}{2}x_i(t_n^-), x_{i+1}(t_n^-), x_{i+2}(t_n^-), \ldots)]\right)$$
$$= \frac{b_i(x_i(t_n^-))}{\varphi(\xi(t_n^-))}$$

for $i \in \mathbb{N}$. In this way, we have checked that the point process $\xi(t)$, $t \geq 0$, is a homogeneous PDMP with values in $\widetilde{X}$.

We can identify the space $\widetilde{X}$ with the space $\mathcal{N}$ of finite counting measures on $\mathbb{R}_+$ by a map $\eta \colon \widetilde{X} \to \mathcal{N}$ given by

$$\eta(\mathbf{x}) = \sum_{\{i \colon x_i \neq *\}} \delta_{x_i} \tag{1.25}$$

where $\mathbf{x} = [(x_1, x_2, \ldots)]$. It means that the process $\eta(\xi(t)), t \geq 0$, is a homogeneous PDMP with values in $\mathcal{N}$.

Remark 1.5 In order to describe the jump transformation at times t_n we need, formally, to introduce a σ-algebra Σ of subset of $\widetilde{X}$ to define a transition probability $P \colon \widetilde{X} \times \Sigma \to [0, 1]$. Usually, Σ is a σ-algebra of Borel subsets of $\widetilde{X}$, thus we need to introduce a topology on the space $\widetilde{X}$. Since the space $\mathcal{N}$ is equipped with the topology of weak convergence of measures, we can define open sets in $\widetilde{X}$ as preimages through the function η of open sets in $\mathcal{N}$. We can also construct directly a metric on the space $\widetilde{X}$. Generally, a point process describes the evolution of a configuration of points in a state space which is a metric space (S, ρ). First, we extend the state space S adding "the dead element" $*$. We need to define a metric on $S \cup \{*\}$. The best situation is if S is a proper subset of a larger metric space S'. Then we simply choose $*$ as an element from S' which does not belong to S and we keep the same metric. In the opposite case, first we choose $x_0 \in S$ and define $\rho(*, x) = 1 + \rho(x_0, x)$ for $x \in S$. Next, we define a metric d on the space X by

$$d(x, y) = \max_{i \in \mathbb{N}} \rho(x_i, y_i)$$

and, finally, we define a metric $\tilde{d}$ on the space $\tilde{X}$ by

$$\tilde{d}([x], [y]) = \min\{d(a, b) : \ a \in [x], \ b \in [y]\}.$$

1.14 Age-Structured Population Model

We consider a population of one sex (usually females) in which all individuals are characterized by their age $x \geq 0$. We assume that an individual with age x dies with rate $d(x)$ and gives birth to one child with rate $b(x)p_1$ or twins with rate $b(x)p_2$, where $p_1 + p_2 = 1$. The model is very similar to that of the size-structured population. We assume that $g(*) = b(*) = d(*) = 0$ and $g(a) = 1$ for $a \in \mathbb{R}_+$. Then the evolution of the population is described by a stochastic process $\xi(t) = [(x_1(t), x_2(t), \ldots)]$ with values in the space $\tilde{X}$ with jumps when an individual dies or gives birth. Between jumps the process $\xi(t)$ satisfies Eq. (1.22) and the jump rate function $\varphi(\mathbf{x})$ is given by (1.23). Let us assume that at the jump time t_{n-1} the population consists of k individuals with ages $x_1, x_2, \ldots, x_k$. It is easy to check that the probability distribution function of $t_{n+1} - t_n$ is given by

$$1 - \exp\left\{ - \int_0^t \sum_{i=1}^k (b(x_i + s) + d(x_i + s)) \, ds \right\}.$$

If $\xi(t_n^-) = \mathbf{x} = [(x_1, x_2, \ldots)]$, then $\xi(t_n)$ has one of the following values:

$$
\begin{aligned}
&[(x_1, \ldots, x_{i-1}, x_{i+1}, \ldots)] \quad \text{with probability } d(x_i)/\varphi(\mathbf{x}) \text{ for } i \in \mathbb{N}, \\
&[(0, x_1, x_2, \ldots)] \quad \text{with probability } b(\mathbf{x})p_1/\varphi(\mathbf{x}), \quad\quad (1.26) \\
&[(0, 0, x_1, x_2, \ldots)] \quad \text{with probability } b(\mathbf{x})p_2/\varphi(\mathbf{x}),
\end{aligned}
$$

where $b(\mathbf{x}) = \sum_{i=1}^{\infty} b(x_i)$.

One can consider more general model when the birth and death rates depend not only on the age of an individual but also on the state of the whole population (e.g. on the size of the population). Let $\mathbf{x} = [(x_1, x_2, \ldots)]$ be the state of the population and $b(x_i, \mathbf{x}), d(x_i, \mathbf{x})$ be the birth and death rates, respectively, of an individual with age

x_i. Even in this case the stochastic process $\xi(t)$ is a homogeneous PDMP. Between jumps, the process $\xi(t)$ satisfies Eq. (1.22). The jump rate function is given by

$$\varphi(\mathbf{x}) = \sum_{i=1}^{\infty} (b(x_i, \mathbf{x}) + d(x_i, \mathbf{x})). \tag{1.27}$$

We also need to replace $b(x_i)$ and $d(x_i)$ by $b(x_i, \mathbf{x})$ and $d(x_i, \mathbf{x})$ in formulas (1.26) for the transition at a jump time.

1.15 Asexual Phenotype Population Model

As in Sects. 1.13 and 1.14 we consider a population of one sex in which all individuals are characterized by their phenotype which is described by a vector x from a set $F \subset \mathbb{R}^d$. The phenotype remains constant during the whole life of an individual. As in the previous model the state space of the population is $\widetilde{X}$. Let $\mathbf{x} = [(x_1, x_2, \ldots)]$ be the state of the population. We assume that an individual with phenotype x_i dies with rate $d(x_i, \mathbf{x})$ and gives birth to one child with rate $b(x_i, \mathbf{x})p_1$ or twins with rate $b(x_i, \mathbf{x})p_2$, where $p_1 + p_2 = 1$. Phenotypes of progeny can differ from the parent's phenotype. We assume that if x is the phenotype of the parent then the progeny phenotype is from a set $B \subset F$ with probability $P(x, B)$, where $P \colon F \times \mathscr{B}(F) \to [0, 1]$ is a transition probability on F and $\mathscr{B}(F)$ is a σ-algebra of Borel subsets of F. We assume that the phenotype of twins is the same (but it is not difficult to consider a model which is based on a different assumption). The point process $\xi(t) = [(x_1(t), x_2(t), \ldots)]$ which describes the phenotype structure of the population is a pure jump-type homogeneous Markov process, i.e. $\xi'(t) = 0$. The jump rate function $\varphi(\mathbf{x})$ is given by (1.27). Since $\xi(t)$ is a pure jump-type homogeneous Markov process the probability distribution function of $t_{n+1} - t_n$ is given by $1 - e^{-\varphi(\mathbf{x})t}$. If $\xi(t_n^-) = \mathbf{x} = [(x_1, x_2, \ldots)]$, then

$$\mathrm{Prob}(\xi(t_n) = [(x_1, \ldots, x_{i-1}, x_{i+1}, \ldots)]) = \frac{d(x_i, \mathbf{x})}{\varphi(\mathbf{x})} \text{ for } i \in \mathbb{N},$$

$$\mathrm{Prob}(\xi(t_n) \in [(x, x_1, x_2, \ldots)] \colon x \in B) = \sum_{i=1}^{\infty} \frac{p_1 b(x_i, \mathbf{x}) P(x_i, B)}{\varphi(\mathbf{x})},$$

$$\mathrm{Prob}(\xi(t_n) \in [(x, x, x_1, x_2, \ldots)] \colon x \in B) = \sum_{i=1}^{\infty} \frac{p_2 b(x_i, \mathbf{x}) P(x_i, B)}{\varphi(\mathbf{x})}.$$

It should be noted that in cellular populations we need to change the last two formulas because progeny consists of two cells ($p_1 = 0$), and we lose a cell which replicates. In this case, we have

$$\mathrm{Prob}(\xi(t_n) \in [(x, x, x_1, \ldots, x_{i-1}, x_{i+1}, \ldots)] \colon x \in B) = \frac{b(x_i, \mathbf{x}) P(x_i, B)}{\varphi(\mathbf{x})}.$$

Also the assumption that both daughter cells have the same phenotype is generally not correct. If we consider hematopoietic stem cells—precursors of blood cells living in the bone marrow—then such cells can be at different levels of morphological development. After division a daughter cell can remain on the same level as the mother cell or go to the next level. We can introduce a model with a point process to describe the evolution of hematopoietic stem cells but a model which is based on a continuous-time Markov chain seems to be more appropriate.

1.16   Phenotype Model with a Sexual Reproduction

Modelling population with sexual reproduction is a challenging task because such models depend on many factors including social behaviour of individuals. The main problem is with the description of the mating process [104]. The simplest case is when each individual has both male and female reproductive organs which is a normal condition in most plants and invertebrates. Then we can consider semi-random or assortative mating. In the case of semi-random mating we assume that an individual with phenotype $x \in F$ has mating rate $p(x)$. It means that if the population consists of k individuals with phenotypes $x_1, x_2, \ldots, x_k$ then individuals with phenotype x_i and x_j form a pair with rate

$$p(x_i, x_j) = \frac{p(x_i)p(x_j)}{\sum_{r=1}^{k} p(x_r)}. \tag{1.28}$$

If we exclude self-fertilization then this rate can be given by

$$p(x_i, x_j) = \frac{1}{2} \frac{p(x_i)p(x_j)}{\sum_{r=1, r \neq i}^{k} p(x_r)} + \frac{1}{2} \frac{p(x_i)p(x_j)}{\sum_{r=1, r \neq j}^{k} p(x_r)} \tag{1.29}$$

for $i \neq j$ and $p(x_i, x_j) = 0$ if $i = j$. We assume that if x_1 and x_2 are the parents' phenotypes then the progeny phenotype is from a set $B \subset F$ with probability $P(x_1, x_2, B)$, where $P : F \times F \times \mathscr{B}(F) \to [0, 1]$ is a transition probability from $F \times F$ to F, i.e. $P(x_1, x_2, \cdot)$ is a probability measure on $\mathscr{B}(F)$ for all x_1, x_2 and for each Borel set $B \subset F$, $(x_1, x_2) \mapsto P(x_1, x_2, B)$ is a measurable function. For simplicity we consider a model when the result of reproduction is only one child. If $\mathbf{x} = [(x_1, x_2, \ldots)]$ is the state of the population then $d(x_i, \mathbf{x})$ is the death rate for an individual with phenotype x_i. As in the previous model the population is described by a pure jump-type homogeneous Markov process $\xi(t) = [(x_1(t), x_2(t), \ldots)]$. The jump rate function $\varphi(\mathbf{x})$ is given by

$$\varphi(\mathbf{x}) = \sum_{i=1}^{\infty} d(x_i, \mathbf{x}) + \sum_{i=1}^{\infty} \sum_{j=1}^{\infty} p(x_i, x_j), \tag{1.30}$$

where $d(*, \mathbf{x}) = 0$ and $p(*, x_i) = p(x_i, *) = 0$ for $i \in \mathbb{N}$. If $\xi(t_n^-) = \mathbf{x} = [(x_1, x_2, \ldots)]$, then

$$\text{Prob}(\xi(t_n) = [(x_1, \ldots, x_{i-1}, x_{i+1}, \ldots)]) = \frac{d(x_i, \mathbf{x})}{\varphi(\mathbf{x})} \text{ for } i \in \mathbb{N},$$

$$\text{Prob}(\xi(t_n) \in [(x, x_1, x_2, \ldots)] : x \in B) = \sum_{i=1}^{\infty} \sum_{j=1}^{\infty} \frac{p(x_i, x_j) P(x_i, x_j, B)}{\varphi(\mathbf{x})}.$$

We now consider assortative mating. Then individuals with similar traits mate more often than they would choose a partner randomly. Then we can use matching theory, according to which each participant ranks all potential partners according to its preferences and attempts to pair with the one with highest-ranking [2], or we can adapt a model based on a preference function [41] used in two-sex populations models. In assortative mating, a preference function $a(x, y)$ is usually of the form $a(x, y) = \psi(\|x - y\|)$, where $\psi : [0, \infty) \to [0, \infty)$ is a continuous and decreasing function. We assume that two individuals with phenotypes x_i and x_j form a pair with rate

$$p(x_i, x_j, \mathbf{x}) = \frac{a(x_i, x_j)}{\sum_{l=1}^{\infty} a(x_i, x_l)} = \frac{\psi(\|x_i - x_j\|)}{\sum_{l=1}^{\infty} \psi(\|x_i - x_l\|)} \tag{1.31}$$

and the rest of the model is similar to that with semi-random mating.

If we consider a two-sex population (i.e. each individual is exclusively male or female) then the mating process is more complex and there are only a few mathematical models of it (see e.g. [60] for insect populations). Since the role of males and females is different, it is clear that in such models we should characterize an individual by its sex and phenotype. An important factor is a sexual selection [57] which often depends on the social behaviour of individuals. Males usually produce enough sperm to inseminate many females and in the case when they do not take part in parental care their reproductive success depends on their phenotype. We give some hints how to build a model in this case. First each individual is described by a pair (x, s), where x is its phenotype, $s = 0$ if it is a male and $s = 1$ if it is a female. Let $\mathbf{x} = [((x_1, s_1), (x_2, s_2), \ldots)]$ be the state of the population. We assume that a female with phenotype x_i gives birth with rate $b(x_i, \mathbf{x})$ and a male with phenotype x_j has the competition rate $p(x_j, \mathbf{x})$. To simplify notation we set $b(x_i, \mathbf{x}) = 0$ if $s_i = 0$ and $p(x_j, \mathbf{x}) = 0$ if $s_i = 1$. A female with phenotype x_i and a male with phenotype x_j form a pair with rate

$$p(x_i, x_j, \mathbf{x}) = \frac{b(x_i, \mathbf{x}) p(x_j, \mathbf{x})}{\sum_{r=1}^{\infty} p(x_r, \mathbf{x})}. \tag{1.32}$$

The rest of the model is similar to the model of hermaphrodite population and we omit it here.

1.17   Coagulation-Fragmentation Process in a Phytoplankton Model

Mathematical modelling of plankton behaviour is a complex issue involving various mathematical tools. A review of mathematical models of plankton dynamics can be found in [102]. Phytoplankton cells tend to form aggregates in which they live together like colonial organisms. Since the size of aggregates is important in the study of fish recruitment, the change of size distribution of aggregates is a very interesting problem both from biological and mathematical point of view. We now introduce a point process which corresponds to a model of phytoplankton dynamics introduced in [6], which takes into account growth and death of aggregates as well as coagulation-fragmentation processes. More advanced models including a space distribution of aggregates and diffusion can be found in [103].

In this model individuals are aggregates and they are characterized by their size $x > 0$, which depends on the number of cells. The division or death of individual cells changes the size of aggregates according to the equation $x'(t) = g(x(t))$ but aggregates can die, for example by sinking to a seabed or whatever cause, with mortality rate $d(x)$. An aggregate can break with rate $b(x)$ and the size y of its descendants is given by the conditional density $k(y, x)$, where k satisfies $k(y, x) = k(x - y, x)$. We have also a coagulation (aggregation) process, by which two distinct aggregates join together to form a single one. We assume that the ability to glue to another aggregate is given by the function $p(x)$. Since the same aggregate cannot join with itself, we assume that two aggregates with sizes x_i and x_j form a new aggregate with rate $p(x_i, x_j)$ given by (1.29) for $i \neq j$ and $p(x_i, x_j) = 0$ if $i = j$.

Our model is similar to that from Sect. 1.13. The population of aggregates is described by a point process $\xi(t)$ with values in $\widetilde{X}$ and satisfies Eq. (1.22). It remains to incorporate to the model fragmentation and coagulation processes. We have a jump if one of aggregates dies, splits or two aggregates form a new one. The jump rate function $\varphi(\mathbf{x})$ at state $\mathbf{x} = [(x_1, x_2, \ldots)]$ is the sum of mortality, fragmentation and coagulation rates:

$$\varphi(\mathbf{x}) = \sum_{i=1}^{\infty} (d(x_i) + b(x_i)) + \sum_{i=1}^{\infty} \sum_{j=1}^{\infty} p(x_i, x_j), \qquad (1.33)$$

where $d(*) = b(*) = p(x_i, *) = p(*, x_i) = 0$ for $i \in \mathbb{N}$. The probability distribution function of $t_{n+1} - t_n$ is given by (1.24). If $\xi(t_n^-) = \mathbf{x} = [(x_1, x_2, \ldots)]$, then

$$\text{Prob}(\xi(t_n) = [(x_1, \ldots, x_{i-1}, x_{i+1}, \ldots)]) = \frac{d(x_i)}{\varphi(\mathbf{x})},$$

$$\text{Prob}(\xi(t_n) \in [(x_1, \ldots, x_{i-1}, x, x_i - x, x_{i+1}, \ldots)]: x \in B) = \frac{b(x_i) \int_B k(y, x_i)\, dy}{\varphi(\mathbf{x})},$$

$$\text{Prob}(\xi(t_n) = [(x_i + x_j, x_1, \ldots, x_{i-1}, x_{i+1}, \ldots, x_{j-1}, x_{j+1}, \ldots)]) = \frac{p(x_i, x_j)}{\varphi(\mathbf{x})}$$

for $i, j \in \mathbb{N}$.

1.18 Paralog Families

We now apply a point process to a model of evolution of paralog families in a genome [98]. Two genes present in the same genome are said to be *paralogs* if they are genetically identical. Let us assume that x_i is a number of copies of a given gene i (we do not repeat paralogous genes). We want to find a point process whose points are sizes of paralog families. The model is based on three fundamental evolutionary events: gene loss, duplication and accumulated change, called for simplicity a mutation. A single gene can:

- be *duplicated* with rate d,
- be *removed* from the genome with rate r,
- *mutate* with rate m and then the gene starts a new one-element paralog family.

Moreover, we assume that all elementary events are independent of each other.

As a result of fundamental evolutionary events a family of size x_i increases to size $x_i + 1$ with rate dx_i, decreases to size $x_i - 1$ with rate rx_i, and decreases to size $x_i - 1$ and forms a new family of size 1 with rate mx_i (Fig. 1.9). We have a jump at any of fundamental evolutionary events. The point process $\xi(t) = [(x_1(t), x_2(t), \ldots)]$ which describes the structure of paralogs is a pure jump-type homogeneous Markov process with values in the space $\widetilde{X} = X/\sim$. In this model we set $* = 0$. Thus X is the space of all sequences with non-negative integer values. The jump rate function $\varphi(\mathbf{x})$ at state $\mathbf{x} = [(x_1, x_2, \ldots)]$ is given by

$$\varphi(\mathbf{x}) = \sum_{i=1}^{\infty} (d + r + m)x_i. \tag{1.34}$$

The probability distribution function of $t_{n+1} - t_n$ is given by $1 - e^{-\varphi(\mathbf{x})t}$, where $\mathbf{x} = \xi(t_n)$. If $\xi(t_n^-) = \mathbf{x} = [(x_1, x_2, \ldots)]$, then

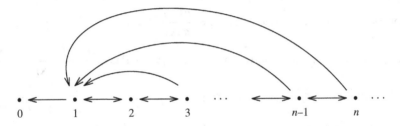

Fig. 1.9 The diagram of the transition between states in the paralog families model

$$\text{Prob}(\xi(t_n) = [(x_1, \ldots, x_{i-1}, x_i + 1, x_{i+1}, \ldots)]) = \frac{dx_i}{\varphi(\mathbf{x})},$$

$$\text{Prob}(\xi(t_n) = [(x_1, \ldots, x_{i-1}, x_i - 1, x_{i+1}, \ldots)]) = \frac{rx_i}{\varphi(\mathbf{x})},$$

$$\text{Prob}(\xi(t_n) = [(1, x_1, \ldots, x_{i-1}, x_i - 1, x_{i+1}, \ldots)]) = \frac{mx_i}{\varphi(\mathbf{x})},$$

for $i \in \mathbb{N}$.

Remark 1.6 The main subject of the paper [98] is the size distribution of paralogous gene families in a genome. Let $s_k(t)$ be the number of k-element families in our model at time t. It means that $s_k(t) = \#\{i : x_i(t) = k\}$. It follows from the description of our model that

$$s_1'(t) = -(d + r)s_1(t) + 2(2m + r)s_2(t) + m \sum_{j=3}^{\infty} js_j(t), \tag{1.35}$$

$$s_k'(t) = d(k - 1)s_{k-1}(t) - (d + r + m)ks_k(t) + (r + m)(k + 1)s_{k+1}(t) \tag{1.36}$$

for $k \geq 2$.

1.19 Definition of PDMP

Having introduced examples of PDMPs, we are ready to give a formal definition of a piecewise deterministic Markov process. In order to do it we need some auxiliary definitions.

Let $\tilde{X}$ be a topological space. A function $\pi : \mathbb{R}_+ \times \tilde{X} \to \tilde{X}$ such that

(a) $\pi(0, x) = x$ for $x \in \tilde{X}$,
(b) $\pi(s + t, x) = \pi(t, \pi(s, x))$ for $x \in \tilde{X}$, $s, t \in \mathbb{R}_+$,
(c) for each $t \geq 0$ the mapping $\pi_t : \tilde{X} \to \tilde{X}$ is continuous with $\pi_t(x) = \pi(t, x)$, $x \in \tilde{X}$,
(d) for each $x \in \tilde{X}$ the mapping $t \mapsto \pi(t, x)$ is càdlàg,, i.e. right-continuous and has left-hand limits

$$\pi(t^+, x) := \lim_{s \downarrow t} \pi(t, x) = \pi(t, x) \quad \text{and} \quad \pi(t^-, x) := \lim_{s \uparrow t} \pi(s, x) \text{ exists in } \tilde{X},$$

is called a *semiflow* or *dynamical system* on $\tilde{X}$.

Given a semiflow on $\tilde{X}$ we define a *survival function* $[0, \infty) \times \tilde{X} \ni (t, x) \mapsto \Phi_x(t) \in [0, 1]$ as a mapping with the following properties:

(a) for each t the transformation $x \mapsto \Phi_x(t)$ is measurable,
(b) for each $x \in \tilde{X}$ the mapping $t \mapsto \Phi_x(t)$ is right-continuous nonincreasing function,
(c) it satisfies

$$\Phi_x(t + s) = \Phi_x(t)\Phi_{\pi(t,x)}(s), \quad t, s \ge 0, x \in \tilde{X}. \tag{1.37}$$

In particular, $1 - \Phi_x$ is the distribution function of a non-negative finite random variable, provided that $\Phi_x(\infty^-) := \lim_{t \to \infty} \Phi_x(t) = 0$. Otherwise, we extend Φ_x to $[0, \infty]$ by setting $\Phi_x(\infty) = 1 - \Phi_x(\infty^-)$. The function Φ defines a probability measure on $[0, \infty]$ by

$$\Phi_x((s, t]) = \Phi_x(s) - \Phi_x(t), \quad t \ge s \ge 0.$$

An example of a survival function is a mapping defined by

$$\Phi_x(t) = \exp\left\{-\int_0^t \varphi(\pi(r, x)) \, dr\right\}, \quad t \ge 0, \tag{1.38}$$

where $\varphi: \tilde{X} \to \mathbb{R}_+$, called a *jump rate function*, is a measurable function such that the function $r \mapsto \varphi(\pi(r, x))$ is integrable on $[0, \varepsilon(x))$ for some $\varepsilon(x) > 0$ and all x. To see that (1.37) holds note that

$$\int_0^{t+s} \varphi(\pi(r, x)) \, dr = \int_0^t \varphi(\pi(r, x)) \, dr + \int_t^{t+s} \varphi(\pi(r, x)) \, dr$$

$$= \int_0^t \varphi(\pi(r, x)) \, dr + \int_0^s \varphi(\pi(r, \pi(t, x))) \, dr.$$

Another example of a survival function is given in Remark 1.4.

Suppose now that X is a Borel subset of $\tilde{X}$. We define the exit time from X by

$$t_*(x) = \sup\{t > 0: \pi(s, x) \in X \text{ for } s \in [0, t)\}, \quad x \in X. \tag{1.39}$$

Let Γ be a subset of the boundary ∂X of X defined by

$$\Gamma = \{\pi(t_*(x)^-, x): 0 < t_*(x) < \infty, \ x \in X\}. \tag{1.40}$$

We call Γ the *active boundary* of the set X. If the active boundary Γ of a set X is non-empty then we can also consider the following survival function

$$\Phi_x(t) = 1_{[0,\, t_*(x))}(t) \exp\left\{ -\int_0^t \varphi(\pi(r, x))\, dr \right\}, \quad t \geq 0, \qquad (1.41)$$

where φ is a jump rate function.

A piecewise deterministic Markov process (PDMP) with values in $X \subset \tilde{X}$ is determined through three *characteristics* (π, Φ, P) which are a semiflow $\pi : \mathbb{R}_+ \times X \to \tilde{X}$, a survival function Φ, and a *jump distribution* $P : (X \cup \Gamma) \times \mathcal{B}(X) \to [0, 1]$ which is a transition probability, i.e. for each set $B \in \mathcal{B}(X)$ the function $x \mapsto P(x, B)$ is measurable and for each $x \in X \cup \Gamma$ the function $B \mapsto P(x, B)$ is a probability measure.

Let us briefly describe the construction of the *piecewise deterministic process* (PDP) $\{\xi(t)\}_{t \geq 0}$ with characteristics (π, Φ, P). We extend the state space X to $X_\Delta = X \cup \{\Delta\}$ where Δ is a fixed state outside X representing a 'dead' state for the process and being an isolated point. For each $x \in X$, we set $P(x, \{\Delta\}) = 0$ and $\pi(t, x) = \Delta$ if $t = \infty$. We also set $\pi(t, \Delta) = \Delta$ for all $t \geq 0$, $P(\Delta, \{\Delta\}) = 1$, and $\Phi_\Delta(t) = 1$ for all $t \geq 0$. Let $(\Omega, \mathscr{F}, \mathbb{P})$ be a probability space. Let $\tau_0 = \sigma_0 = 0$ and let $\xi(0) = \xi_0$ be an X-valued random variable. For each $n \geq 1$, we can choose a positive random variable σ_n satisfying

$$\mathbb{P}(\sigma_n > t | \xi_{n-1} = x) = \Phi_x(t), \quad t \in [0, \infty].$$

Define the nth *jump time* by

$$\tau_n = \tau_{n-1} + \sigma_n$$

and set

$$\xi(t) = \begin{cases} \pi(t - \tau_{n-1}, \xi_{n-1}) & \text{for } \tau_{n-1} \leq t < \tau_n, \\ \xi_n & \text{for } t = \tau_n, \end{cases}$$

where the nth *post-jump position* ξ_n is an X_Δ-valued random variable such that

$$\mathbb{P}(\xi_n \in B | \xi(\tau_n^-) = x) = P(x, B), \quad x \in X_\Delta \cup \Gamma,$$

and $\xi(\tau_n^-) = \lim_{t \uparrow \tau_n} \xi(t)$. Thus, the trajectory of the process is defined for all $t < \tau_\infty := \lim_{n \to \infty} \tau_n$ and τ_∞ is called the *explosion time*. To define the process for all times, we set $\xi(t) = \Delta$ for $t \geq \tau_\infty$. The process $\{\xi(t)\}_{t \geq 0}$ is called the *minimal* PDP corresponding to (π, Φ, P). It has right-continuous sample paths, by construction. In the next chapter, we will show that the minimal PDP exists and is a Markov process.

Chapter 2
Markov Processes

In this chapter, we provide a background material that is needed to define and study Markov processes in discrete and continuous time. We start by giving basic examples of transition probabilities and the corresponding operators on the spaces of functions and measures. An emphasis is put on stochastic operators on the spaces of integrable functions. The importance of transition probabilities is that the distribution of a stochastic process with Markov property is completely determined by transition probabilities and initial distributions. The Markov property simply states that the past and the future are independent given the present. We refer the reader to Appendix A for the required theory on measure, integration, and basic concepts of probability theory.

2.1 Transition Probabilities and Kernels

2.1.1 Basic Concepts

In this section, we introduce transition kernels, stressing the interplay between analytic and stochastic interpretations. Let X be a set and let Σ be a σ-algebra of subsets of X. The pair (X, Σ) is called a measurable space. A function $P\colon X \times \Sigma \to [0, 1]$ is said to be a *transition kernel* if

(1) for each set $B \in \Sigma$, $P(\cdot, B)\colon X \to [0, 1]$ is a measurable function;
(2) for each $x \in X$ the set function $P(x, \cdot)\colon \Sigma \to [0, 1]$ is a measure.

If $P(x, X) = 1$ for all $x \in X$ then P is said to be a *transition probability* on (X, Σ).

Let $(\Omega, \mathscr{F}, \mathbb{P})$ be a probability space. A mapping $\xi\colon \Omega \to X$ is said to be an *X-valued random variable* if it is measurable, i.e. for any $B \in \Sigma$

© The Author(s) 2017
R. Rudnicki and M. Tyran-Kamińska, *Piecewise Deterministic Processes in Biological Models*, SpringerBriefs in Mathematical Methods, DOI 10.1007/978-3-319-61295-9_2

$$\{\xi \in B\} = \xi^{-1}(B) = \{\omega \in \Omega : \xi(\omega) \in B\} \in \mathscr{F}.$$

The *(probability) distribution* or the *(probability) law* of ξ is a probability measure μ_ξ on (X, Σ) defined by

$$\mu_\xi(B) = \mathbb{P}(\xi \in B), \quad B \in \Sigma.$$

Given an X-valued random variable ξ on the probability space $(\Omega, \mathscr{F}, \mathbb{P})$ the family

$$\sigma(\xi) = \{\xi^{-1}(B) : B \in \Sigma\}$$

is a σ-algebra of subsets of Ω and it is called the *σ-algebra generated by ξ*. Note that ξ is an X-valued random variable if and only if $\sigma(\xi) \subseteq \mathscr{F}$. If $\mathscr{G}$ is a sub-σ-algebra of $\mathscr{F}$ then ξ is said to be $\mathscr{G}$-measurable if $\sigma(\xi) \subseteq \mathscr{G}$. Thus one can also say that $\sigma(\xi)$ is the smallest σ-algebra $\mathscr{G}$ such that ξ is $\mathscr{G}$-measurable. Recall that sub-σ-algebras $\mathscr{F}_1$ and $\mathscr{F}_2$ of $\mathscr{F}$ are called *independent* if $\mathbb{P}(A_1 \cap A_2) = \mathbb{P}(A_1)\mathbb{P}(A_2)$ for any $A_1 \in \mathscr{F}_1$ and $A_2 \in \mathscr{F}_2$. If a random variable ϑ with values in any measurable space is *independent of the σ-algebra $\mathscr{G}$*, i.e. if $\sigma(\vartheta)$ and $\mathscr{G}$ are independent, and if ξ is an $\mathscr{G}$-measurable X-valued random variable, then the conditional expectation of $g(\xi, \vartheta)$ with respect to $\mathscr{G}$ is given by

$$\mathbb{E}(g(\xi, \vartheta)|\mathscr{G}) = \mathbb{E}(g(\xi, \vartheta)|\xi) \quad \text{and} \quad \mathbb{E}(g(\xi, \vartheta)|\xi = x) = \mathbb{E}(g(x, \vartheta)) \qquad (2.1)$$

for all $x \in X$ and for any measurable non-negative function g; see Lemma A.3.

Let Y be a metric space. Consider a probability measure ν on the Borel σ-algebra $\mathscr{B}(Y)$. Given a measurable mapping $\kappa : X \times Y \to X$, we define

$$P(x, B) = \int \mathbf{1}_B(\kappa(x, y))\nu(dy), \quad x \in X, \ B \in \Sigma, \qquad (2.2)$$

where $\mathbf{1}_B$ is the indicator function, equal to one on the set B and zero otherwise. Clearly, P is a transition probability. For a Y-valued random variable ϑ with distribution ν and for every $x \in X$, the X-valued random variable $\kappa(x, \vartheta)$ has distribution $P(x, \cdot)$, since

$$\mathbb{P}(\kappa(x, \vartheta) \in B) = \mathbb{E}(\mathbf{1}_B(\kappa(x, \vartheta))) = \int \mathbf{1}_B(\kappa(x, y))\nu(dy) = P(x, B), \quad B \in \Sigma.$$

As a particular example, we can take Y to be the unit interval $[0, 1]$ and ν to be the Lebesgue measure on $[0, 1]$; in that case ϑ is said to be uniformly distributed on the unit interval $[0, 1]$. Now let ξ_0 be an X-valued random variable independent of ϑ and let $\xi_1 = \kappa(\xi_0, \vartheta)$. We can write

$$\mathbb{P}(\xi_1 \in B|\xi_0 = x) = \mathbb{P}(\kappa(\xi_0, \vartheta) \in B|\xi_0 = x) = \mathbb{E}(\mathbf{1}_B(\kappa(\xi_0, \vartheta))|\xi_0 = x)$$

and if we take $g(x, y) = \mathbf{1}_B(\kappa(x, y))$ in (2.1) then $\mathbb{E}(g(x, \vartheta)) = P(x, B)$. Thus

$$\mathbb{P}(\xi_1 \in B | \xi_0 = x) = P(x, B),$$

which gives the *conditional distribution* of ξ_1 given ξ_0. Moreover, if μ_{ξ_0} is the distribution of ξ_0 then the joint distribution $\mu_{(\xi_0, \xi_1)}$ of ξ_0 and ξ_1 satisfies

$$\mu_{(\xi_0, \xi_1)}(B_0 \times B_1) = \mathbb{P}(\xi_0 \in B_0, \xi_1 \in B_1) = \int_{B_0} P(x, B_1) \mu_{\xi_0}(dx)$$

for all $B_0, B_1 \in \Sigma$.

We need an extension of the concept of transition probabilities. Let (X_0, Σ_0) and (X_1, Σ_1) be two measurable spaces. A function $P : X_0 \times \Sigma_1 \to [0, 1]$ is said to be a *transition probability from* (X_0, Σ_0) to (X_1, Σ_1) if $P(\cdot, B)$ is measurable for every $B \in \Sigma_1$ and $P(x, \cdot)$ is a probability measure on (X_1, Σ_1) for every $x \in X_0$. The next result ensures existence of probability measures on product spaces.

Proposition 2.1 *Let P be a transition probability from (X_0, Σ_0) to (X_1, Σ_1). If μ_0 is a probability measure on (X_0, Σ_0) then there exists a unique probability measure μ on the product space $(X_0 \times X_1, \Sigma_0 \otimes \Sigma_1)$ such that*

$$\mu(B_0 \times B_1) = \int_{B_0} P(x_0, B_1) \mu_0(dx_0), \quad B_0 \in \Sigma_0, B_1 \in \Sigma_1.$$

Proof Let $B^{x_0} = \{x_1 \in X_1 : (x_0, x_1) \in B\}$ for every $x_0 \in X_0$ and $B \in \Sigma_0 \otimes \Sigma_1$. Clearly, if $B = B_0 \times B_1$ then we have $B^{x_0} = B_1$ for $x_0 \in B_0$ and $B^{x_0} = \emptyset$ for $x_0 \notin B_0$. By the monotone class theorem (see Theorem A.1) for each $B \in \Sigma_0 \otimes \Sigma_1$ the function $x_0 \mapsto P(x_0, B^{x_0})$ is measurable. Define

$$\mu(B) = \int_{X_0} P(x_0, B^{x_0}) \mu_0(dx_0), \quad B \in \Sigma_0 \otimes \Sigma_1,$$

which is clearly a measure on the product space. Since the product σ-algebra is generated by rectangles, i.e. $\Sigma_0 \otimes \Sigma_1 = \sigma(\mathscr{C})$, where $\mathscr{C} = \{B_0 \times B_1 : B_0 \in \Sigma_0, B_1 \in \Sigma_1\}$, and $\mathscr{C}$ is a π-system, the result follows from the $\pi - \lambda$ lemma. $\qquad\qquad\qquad\square$

We now show that, for a large class of measurable spaces, any transition probability is as in (2.2). We call a measurable space (X, Σ) a *Borel space* if X is isomorphic to a Borel subset of $[0, 1]$, i.e. there exists a Borel subset Y of $[0, 1]$ and a measurable bijection $\psi : X \to Y$ such that its inverse ψ^{-1} is also measurable. Any Borel subset of a *Polish space*, i.e. a complete separable metric space, with $\Sigma = \mathscr{B}(X)$ is a Borel space (see e.g. [55, Theorem A1.2]). Recall that a metric space (X, ρ) is *complete* if every sequence (x_n) satisfying the Cauchy condition $\lim_{m,n \to \infty} \rho(x_n, x_m) = 0$ is convergent to some $x \in X$: $\lim_{n \to \infty} \rho(x_n, x) = 0$, and it is *separable* if there is a countable set $X_0 \subset X$ which is dense in X, so that for any $x \in X$ there exists a sequence from X_0 converging to x.

Proposition 2.2 *Let P be a transition probability from a measurable space* (X_0, Σ_0) *to a Borel space* (X_1, Σ_1). *Then there exists a measurable function* $\kappa : X_0 \times [0, 1] \to X_1$ *such that if a random variable* ϑ *is uniformly distributed on the unit interval* $[0, 1]$ *then* $\kappa(x, \vartheta)$ *has distribution* $P(x, \cdot)$ *for every* $x \in X_0$.

Proof Since $\psi : X_1 \to \psi(X_1)$ with $\psi(X_1) \subset [0, 1]$ is a bijection and we have $\psi^{-1}([0, t]) = \psi^{-1}([0, t] \cap \psi(X_1))$, we can define

$$F_x(t) = P(x, \psi^{-1}([0, t])), \quad t \in (0, 1].$$

For each x the function $t \mapsto F_x(t)$ is right-continuous with $F_x(1) = 1$. Consider its generalized inverse

$$F_x^{\leftarrow}(q) := \inf\{t \geq 0 : F_x(t) \geq q\}, \quad q \geq 0,$$

with the convention that the infimum of an empty set is equal to $+\infty$, and define $\kappa(x, q) = \psi^{-1}(F_x^{\leftarrow}(q))$ if $F_x^{\leftarrow}(q) \in \psi(X_1)$ and $\kappa(x, q) = x_1$ if $F_x^{\leftarrow}(q) \notin \psi(X_1)$, where x_1 is an arbitrary point from X_1. We have

$$\mathrm{Leb}\{q \in [0, 1] : \kappa(x, q) \in B\} = P(x, B) \quad \text{for } x \in X_0, \ B \in \Sigma_1,$$

where Leb denotes the Lebesgue measure on $[0, 1]$.

2.1.2   Transition Operators

It is convenient to associate with each transition kernel two linear mappings acting in two Banach spaces, one which is a space of bounded functions and the other which is a space of measures.

Let P be a transition kernel on a measurable space (X, Σ). Consider first the Banach space $B(X)$ of bounded, measurable, real-valued functions on X with the supremum norm

$$\|g\|_u = \sup_{x \in X} |g(x)|.$$

Given $g \in B(X)$ we define

$$Tg(x) = \int_X g(y) P(x, dy), \quad x \in X.$$

If $B \in \Sigma$ then the indicator function $\mathbf{1}_B \in B(X)$ and

$$T\mathbf{1}_B(x) = P(x, B), \quad x \in X.$$

Thus $T1_B \in B(X)$. By the monotone class theorem, we obtain $Tg \in B(X)$ for $g \in B(X)$. From linearity of the integral it follows that the mapping $B(X) \ni g \mapsto Tg \in B(X)$ is a linear operator and that

$$\|Tg\|_u \leq \|g\|_u, \quad g \in B(X).$$

Now, let $\mathcal{M}^+(X)$ be the collection of all finite measures on Σ. For each $\mu \in \mathcal{M}^+(X)$ we define

$$P\mu(B) = \int_X P(x, B)\,\mu(dx), \quad B \in \Sigma.$$

Then $P\mu$ is a finite measure and $P\mu(X) \leq \mu(X)$. The space $\mathcal{M}^+(X)$ is a *cone*, i.e. $c_1\mu_1 + c_2\mu_2 \in \mathcal{M}^+(X)$, if c_1, c_2 are non-negative constants and $\mu_1, \mu_2 \in \mathcal{M}^+(X)$, and we have

$$P(c_1\mu_1 + c_2\mu_2) = c_1 P\mu_1 + c_2 P\mu_2.$$

Therefore we can extend P so that it is a linear operator on the vector space of all finite signed measures

$$\mathcal{M}(X) = \{\mu_1 - \mu_2 : \mu_1, \mu_2 \in \mathcal{M}^+(X)\},$$

by setting $P\mu = P\mu_1 - P\mu_2$ for $\mu = \mu_1 - \mu_2 \in \mathcal{M}(X)$. Define the *total variation norm* on $\mathcal{M}(X)$ by

$$\|\mu\|_{TV} = |\mu|(X), \quad \mu \in \mathcal{M}(X),$$

where the *total variation* of a finite signed measure μ is given by

$$|\mu|(X) = \sup_{B \in \Sigma} \mu(B) - \inf_{B \in \Sigma} \mu(B) = \sup\{|\int g\,d\mu| : g \in B(X),\ \|g\|_u \leq 1\}.$$

Then $(\mathcal{M}(X), \|\cdot\|_{TV})$ is a Banach space and we have

$$\|P\mu\|_{TV} \leq \|\mu\|_{TV}, \quad \mu \in \mathcal{M}(X).$$

In particular, a transition probability P provides the following stochastic interpretation. If X-valued random variables ξ_0 and ξ_1 are such that

$$\mathbb{P}(\xi_1 \in B | \xi_0 = x) = P(x, B)$$

and ξ_0 has a distribution μ_0, then $\mu_1 = P\mu_0$ is the distribution of ξ_1 and

$$\mathbb{E}(g(\xi_1) | \xi_0 = x) = \int_X g(y) P(x, dy) = Tg(x), \quad x \in X,\ g \in B(X).$$

2.1.3 Substochastic and Stochastic Operators

In this section, we suppose that a measurable space (X, Σ) carries a σ-finite measure m. The set of all measurable and m-integrable functions is a linear space and it becomes a Banach space, written $L^1(X, \Sigma, m)$, by defining the norm

$$\|f\| = \int_X |f(x)|\, m(dx), \quad f \in L^1(X, \Sigma, m),$$

and by identifying functions that are equal almost everywhere. Given a measurable function $f: X \to \mathbb{R}$ the *essential supremum* of f is defined by

$$\text{ess sup} |f| = \inf\{c > 0: \ m\{x: |f(x)| > c\} = 0\} =: \|f\|_\infty.$$

The set of all functions with a finite essential supremum is denoted by $L^\infty(X, \Sigma, m)$ and it becomes a Banach space when we identify functions that are equal almost everywhere and take $\| \cdot \|_\infty$ as the norm.

A linear operator $P: L^1(X, \Sigma, m) \to L^1(X, \Sigma, m)$ is called *substochastic* if it is *positive*, i.e. if $f \geq 0$ then $Pf \geq 0$ for $f \in L^1(X, \Sigma, m)$, and if $\|Pf\| \leq \|f\|$ for $f \in L^1(X, \Sigma, m)$. A substochastic operator is called *stochastic* if $\|Pf\| = \|f\|$ for $f \in L^1(X, \Sigma, m)$ and $f \geq 0$. We denote by $P^*: L^\infty(X, \Sigma, m) \to L^\infty(X, \Sigma, m)$ the *adjoint operator* of P, i.e. for every $f \in L^1(X, \Sigma, m)$ and $g \in L^\infty(X, \Sigma, m)$

$$\int_X Pf(x)g(x)\, m(dx) = \int_X f(x)P^*g(x)\, m(dx).$$

Let P be a transition kernel on (X, Σ). Take $f \in L^1(X, \Sigma, m)$ with $f \geq 0$ and $\|f\| = 1$. The measure $\mu \in \mathcal{M}^+(X)$ given by $d\mu = f\, dm$, i.e.

$$\mu(B) = \int_B f(x)\, m(dx), \quad B \in \Sigma,$$

is absolutely continuous with respect to m and we call f the *density* of μ (with respect to m). If the measure $\nu = P\mu$ given by

$$\nu(B) = P\mu(B) = \int_X P(x, B)f(x)\, m(dx), \quad B \in \Sigma,$$

is absolutely continuous with respect to m then, by the Radon–Nikodym theorem, there is an essentially unique $g \in L^1$, $g \geq 0$, such that

$$\nu(B) = \int_B g(x)\, m(dx), \quad B \in \Sigma.$$

The function g is the Radon–Nikodym derivative dv/dm of the measure v with respect to m. If P is a transition probability then we have $P\mu(X) = 1$. Thus f is mapped to g, and we can write

$$Pf = g, \quad \text{if } f = \frac{d\mu}{dm},$$

which is the density of the measure $v = P\mu$

$$P\mu(B) = \int_B Pf(x) \, m(dx).$$

This motivates the following definitions.

Denote by D the subset of $L^1 = L^1(X, \Sigma, m)$ which contains all *densities*, i.e.

$$D = \{f \in L^1 : f \geq 0, \ \|f\| = 1\}.$$

Observe that a linear operator P on $L^1(X, \Sigma, m)$ is a stochastic operator if and only if $P(D) \subseteq D$. The transition kernel P corresponds to a substochastic operator P if

$$\int_X P(x, B) f(x) \, m(dx) = \int_B Pf(x) \, m(dx) \quad \text{for all } B \in \Sigma, \ f \in D,$$

or, equivalently, the adjoint of P is of the form

$$P^* g(x) = \int g(y) \, P(x, dy) \quad \text{for } g \in L^\infty(X, \Sigma, m). \tag{2.3}$$

On the other hand, if a transition kernel P has the following property

$$m(B) = 0 \Longrightarrow P(x, B) = 0 \quad \text{for } m\text{-a.e. } x \text{ and } B \in \Sigma, \tag{2.4}$$

then there exists a substochastic operator P such that (2.3) holds.

Remark 2.1 There exists a stochastic operator which does not have a transition kernel [32]. But if X is a Polish space (a complete separable metric space) and Σ is a σ-algebra of Borel subsets of X, then each substochastic operator on $L^1(X, \Sigma, m)$ has a transition kernel [48]. The same result holds if X is a Borel subset of a Polish space and Σ is the σ-algebra of Borel subsets of X.

2.1.4 Integral Stochastic Operators

Let (X, Σ, m) be a σ-finite measure space. Any measurable function $k : X \times X \to [0, \infty)$ such that

$$\int_X k(x, y)\, m(dx) = 1 \quad \text{for all } y \in X$$

defines a transition probability by

$$P(x, B) = \int_B k(y, x)\, m(dy), \quad B \in \Sigma,$$

called an *integral kernel*. If we set

$$Pf(x) = \int_X k(x, y) f(y)\, m(dy), \quad f \in L^1(X, \Sigma, m),$$

then for $f \geq 0$ we have $Pf \geq 0$ and

$$\int_X Pf(x)\, m(dx) = \int_X \int_X k(x, y)\, m(dx) f(y)\, m(dy) = \int_X f(y)\, m(dy),$$

which shows that P is a stochastic operator. If, instead, k is such that

$$\int_X k(x, y)\, m(dx) \leq 1 \quad \text{for all } y \in X,$$

then $P(x, B)$ is a transition kernel. It defines a substochastic operator on $L^1(X, \Sigma, m)$, called an integral operator.

Suppose that the state space is *discrete*, i.e. the set X is finite or countable and Σ is the family of all subsets of X. Since any probability measure on Σ is uniquely defined by its values on the singleton sets $\{y\}$, $y \in X$, any transition probability P satisfies

$$P(x, B) = \sum_{y \in B} P(x, \{y\}), \quad x \in X, \ B \subseteq X.$$

If the state space is discrete and the measure m is the counting measure then (X, Σ, m) is a σ-finite measure space and every kernel is an integral kernel.

We now consider the case of $X = \mathbb{N} = \{0, 1, \ldots\}$, where we use the notation $l^1 = L^1(X, \Sigma, m)$ with m being the counting measure on X. We represent any function f as a sequence $u = (u_i)_{i \in \mathbb{N}}$. We have $(u_i)_{i \in \mathbb{N}} \in l^1$ if and only if $\sum_{i=0}^{\infty} |u_i| < \infty$. If

$$\sum_{i=0}^{\infty} k_{ij} = 1 \quad \text{for all } j,$$

then the operator $P : l^1 \to l^1$ defined by

$$(Pu)_i = \sum_{j=0}^{\infty} k_{ij} u_j, \quad i \in \mathbb{N},$$

is a stochastic operator. It can be identified with a matrix $P = [k_{ij}]_{i,j \in \mathbb{N}}$, called a *stochastic matrix*. It has non-negative elements and the elements in each column sum up to one. The transposed matrix

$$P^* = [k_{ji}]_{i,j \in \mathbb{N}}, \quad k_{ji} = P(i, \{j\}),$$

is called a *transition matrix* and it is such that the elements in each row sum up to one. Note that k_{ji} is the probability of going from state i to state j.

2.1.5 Frobenius–Perron Operator

Consider a measurable transformation $S: X \to X$, where (X, Σ, m) is a space with a σ-finite measure m. Let μ be a probability measure on (X, Σ) and let us observe the evolution of this measure under the action of the system. For example, if we start with the probability measure concentrated at the point x, i.e. the Dirac measure δ_x, then under the action of the system we obtain the measure $\delta_{S(x)}$. In general, if a measure μ describes the distribution of points in the phase space X, then the measure ν given by the formula $\nu(B) = \mu(S^{-1}(B))$ describes the distribution of points after the action of the transformation S. Assume that μ is absolutely continuous with respect to m with density f. If the measure ν is also absolutely continuous with respect to m, and $g = d\nu/dm$, then we define an operator P_S by $P_S f = g$. This operator corresponds to the transition probability function $P(x, B)$ on (X, Σ) given by

$$P(x, B) = \begin{cases} 1, & \text{if } S(x) \in B, \\ 0, & \text{if } S(x) \notin B. \end{cases} \quad (2.5)$$

The operator P_S is correctly defined if the transition probability function $P(x, B)$ satisfies condition (2.4). Condition (2.4) now takes the form

$$m(B) = 0 \Longrightarrow m(S^{-1}(B)) = 0 \text{ for } B \in \Sigma \quad (2.6)$$

and the transformation S which satisfies (2.6) is called *non-singular*. This operator can be extended to a bounded linear operator $P_S: L^1 \to L^1$, and P_S is a stochastic operator. The operator P_S is called the *Frobenius–Perron operator* for the transformation S or the *transfer operator* or the *Ruelle operator*.

We now give a formal definition of the Frobenius–Perron operator. Let (X, Σ, m) be a σ-finite measure space and let S be a measurable non-singular transformation of X. An operator $P_S: L^1 \to L^1$ which satisfies the following condition

$$\int_B P_S f(x) \, m(dx) = \int_{S^{-1}(B)} f(x) \, m(dx) \quad \text{for } B \in \Sigma \text{ and } f \in L^1 \quad (2.7)$$

is the Frobenius–Perron operator for the transformation S. The adjoint of the
Frobenius–Perron operator $P_S^*: L^\infty \to L^\infty$ is given by $P_S^* g(x) = g(S(x))$ and
is called the *Koopman operator* or the *composition operator*. In particular, if
$S: X \to X$ is one to one and non-singular with respect to m, then

$$P_S f(x) = 1_{S(X)}(x) f(S^{-1}(x)) \frac{d(m \circ S^{-1})}{dm}(x) \quad \text{for } m\text{-a.e. } x \in X,$$

where $d(m \circ S^{-1})/dm$ is the Radon–Nikodym derivative of the measure $m \circ S^{-1}$
with respect to m.

We next show how to find the Frobenius–Perron operator for piecewise smooth
transformations of subsets of $\mathbb{R}^d$. Let X be a subset of $\mathbb{R}^d$ with non-empty interior
and with the boundary of zero Lebesgue measure. Let $S: X \to X$ be a measurable
transformation. We assume that there exist pairwise disjoint open subsets $U_1,\dots,U_n$
of X having the following properties:

(a) the sets $X_0 = X \setminus \bigcup_{i=1}^{n} U_i$ and $S(X_0)$ have zero Lebesgue measure,
(b) maps $S_i = S\big|_{U_i}$ are diffeomorphisms from U_i onto $S(U_i)$, i.e. S_i are C^1 and
     invertible transformations with $\det S_i'(x) \neq 0$ at each point $x \in U_i$.

Then transformations $\psi_i = S_i^{-1}$ are also diffeomorphisms from $S(U_i)$ onto U_i and
the Frobenius–Perron operator P_S exists and is given by the formula

$$P_S f(x) = \sum_{i \in I_x} f(\psi_i(x)) |\det \psi_i'(x)|, \tag{2.8}$$

where $I_x = \{i : \psi_i(x) \in U_i\}$. Indeed,

$$\int_{S^{-1}(B)} f(x)\,dx = \sum_{i=1}^{n} \int_{S^{-1}(B) \cap U_i} f(x)\,dx = \sum_{i=1}^{n} \int_{\psi_i(B)} f(x)\,dx$$

$$= \sum_{i=1}^{n} \int_{B \cap S(U_i)} f(\psi_i(x)) |\det \psi_i'(x)|\,dx$$

$$= \int_B \sum_{i \in I_x} f(\psi_i(x)) |\det \psi_i'(x)|\,dx = \int_B P f(x)\,dx.$$

2.1.6 Iterated Function Systems

We now consider a finite set of different non-singular transformations $S_1, \dots, S_k$ of a
σ-finite measure space (X, Σ, m). Let $p_1(x), \dots, p_k(x)$ be non-negative measurable
functions defined on X such that $p_1(x) + \cdots + p_k(x) = 1$ for all $x \in X$. Take a point
x. We choose a transformation S_j with probability $p_j(x)$ and the position of x after

the action of the system is given by $S_j(x)$. Thus we consider the following transition probability

$$P(x, B) = \sum_{j=1}^{k} p_j(x)\delta_{S_j(x)}(B)$$

for $x \in X$ and measurable sets B. Hence, for any measure μ, we have

$$P\mu(B) = \sum_{j=1}^{k} \int_X p_j(x)\delta_{S_j(x)}(B)\mu(dx) = \sum_{j=1}^{k} \int_{S_j^{-1}(B)} p_j(x)\mu(dx).$$

If μ is absolutely continuous and $f = d\mu/dm$ then

$$P\mu(B) = \sum_{j=1}^{k} \int_{S_j^{-1}(B)} p_j(x)f(x)\,m(dx) = \sum_{j=1}^{k} \int_B P_{S_j}(p_j f)(x)\,m(dx),$$

where $P_{S_1}, \ldots, P_{S_k}$ are the corresponding Frobenius–Perron operators. Consequently, the stochastic operator on $L^1 = L^1(X, \Sigma, m)$ corresponding to P is of the form

$$Pf = \sum_{j=1}^{k} P_{S_j}(p_j f), \quad f \in L^1.$$

We can extend the iterated function system in the following way. Consider a family of measurable transformations $S_y : X \to X$, $y \in Y$, where Y is a metric space which carries a Borel measure v, and a family of measurable functions $p_y : X \to [0, \infty)$, $y \in Y$, satisfying

$$\int_Y p_y(x)v(dy) = 1, \quad x \in X,$$

so that the transition probability P is of the form

$$P(x, B) = \int_Y 1_B(S_y(x))p_y(x)v(dy), \quad x \in X. \tag{2.9}$$

If each S_y is a non-singular transformation of the space (X, Σ, m) then the stochastic operator P corresponding to the transition probability as in (2.9) is of the form

$$Pf = \int_Y P_{S_y}(p_y f)v(dy), \quad f \in L^1,$$

where P_{S_y} is the Frobenius–Perron operator for S_y, $y \in Y$.

As a final example, we look at the transition probability describing the switching mechanism in Sects. 1.8 and 1.9. Let I be at most a countable set. We define the transformation $S_j : X \times I \to X \times I$, $j \in I$, by

$$S_j(x, i) = (x, j), \quad x \in X, \ i, j \in I.$$

Each transformation defined on $\mathbb{X} = X \times I$ is non-singular with respect to the product measure m of a measure on X and the counting measure on I. We assume that $q_{ji}(x)$, $j \neq i$, are non-negative measurable functions satisfying $\sum_{j \neq i} q_{ji}(x) < \infty$ for all $i \in I$, $x \in X$. Then we define the jump rate function by

$$q_i(x) = \sum_{j \neq i} q_{ji}(x)$$

and the probabilities p_j, $j \in I$, by $p_i(x, i) = 0$ and

$$p_j(x, i) = \begin{cases} 1, & q_i(x) = 0, \ j \neq i, \\ \frac{q_{ji}(x)}{q_i(x)}, & q_i(x) > 0, \ j \neq i. \end{cases}$$

The stochastic operator P on $L^1 = L^1(\mathbb{X}, \Sigma, m)$ corresponding to the transition probability

$$P((x, i), \{(x, j)\}) = p_j(x, i)$$

is given by

$$Pf(x, i) = \sum_{j \neq i} p_i(x, j) f(x, j).$$

In particular, if $I = \{0, 1\}$ and $q_i(x) > 0$ for $x \in X$, $i = 0, 1$, then

$$p_j(x, i) = \begin{cases} 0, & j = i, \\ 1, & j = 1 - i, \end{cases}$$

and $Pf(x, i) = P_{S_{1-i}}(x, i) = f(x, 1 - i)$ for $(x, i) \in \mathbb{X} = X \times \{0, 1\}$.

2.2 Discrete-Time Markov Processes

2.2.1 Markov Processes and Transition Probabilities

A family ξ_n, $n \in \mathbb{N}$, of X-valued random variables defined on a probability space $(\Omega, \mathscr{F}, \mathbb{P})$ is called a *discrete-time stochastic process* with *state space* (X, Σ). It is said to have the (weak) *Markov property* if for each time $n \geq 0$

$$\mathbb{P}(\xi_{n+1} \in B | \xi_0, \dots, \xi_n) = \mathbb{P}(\xi_{n+1} \in B | \xi_n), \quad B \in \Sigma. \tag{2.10}$$

The sequence $\mathscr{F}_n^\xi = \sigma(\xi_0, \ldots, \xi_n)$, $n \geq 0$, is called the *natural filtration* or the *history* of the process (ξ_n). We have

$$\sigma(\xi_0, \ldots, \xi_n) = \{\{\xi_0 \in B_0, \ldots, \xi_n \in B_n\} : B_0, \ldots, B_n \in \Sigma\}.$$

The monotone class theorem and properties of the conditional expectation imply that if (ξ_n) has the weak Markov property then for each $k \geq 0$ and all $g_0, \ldots, g_k \in B(X)$ the following holds

$$\mathbb{E}(g_0(\xi_n) \ldots g_k(\xi_{n+k}) | \xi_0, \ldots, \xi_n) = \mathbb{E}(g_0(\xi_n) \ldots g_k(\xi_{n+k}) | \xi_n).$$

We can extend it further as follows. A sequence (ξ_n) has the weak Markov property if and only if the future $\sigma(\xi_m : m \geq n)$ and the past $\sigma(\xi_m : m \leq n)$ are conditionally independent given the present $\sigma(\xi_n)$ (see Lemma A.4), i.e. for each n and all $A \in \sigma(\xi_m : m \geq n)$ and $F \in \sigma(\xi_m : m \leq n)$ we have

$$\mathbb{P}(A \cap F | \xi_n) = \mathbb{P}(A | \xi_n) \mathbb{P}(F | \xi_n).$$

A discrete-time stochastic process $\xi = (\xi_n)_{n \geq 0}$ is called a *(homogeneous) Markov process* with transition probability P and initial distribution μ on (X, Σ) if there is a probability space $(\Omega, \mathscr{F}, \mathbb{P})$ such that

$$\mu(B) = \mathbb{P}(\xi_0 \in B) \quad \text{and} \quad \mathbb{P}(\xi_{n+1} \in B | \xi_0, \ldots, \xi_n) = P(\xi_n, B) \tag{2.11}$$

for all $B \in \Sigma$ and $n \geq 0$. Taking the conditional expectation of (2.11) with respect to $\sigma(\xi_n)$ we obtain

$$\mathbb{P}(\xi_{n+1} \in B | \xi_n) = P(\xi_n, B)$$

which implies that $\xi = (\xi_n)_{n \geq 0}$ has the weak Markov property. Moreover, we have

$$\mathbb{P}(\xi_{n+k} \in B | \xi_0, \ldots, \xi_n) = \mathbb{P}(\xi_{n+k} \in B | \xi_n) = P^k(\xi_n, B), \quad k \geq 1,$$

where P^k, called the *kth step transition probability*, is defined inductively by

$$P^0(x, B) = \delta_x(B), \quad P^{k+1}(x, B) = \int_X P(y, B) \, P^k(x, dy).$$

The kernels P^k, $k \geq 1$, satisfy the *Chapman–Kolmogorov equation*

$$P^{k+n}(x, B) = \int_X P^k(y, B) \, P^n(x, dy). \tag{2.12}$$

We can regard the stochastic process ξ_n, $n \geq 0$, as an $X^{\mathbb{N}}$-valued random variable, where $X^{\mathbb{N}}$ is the product space

$$X^{\mathbb{N}} = \{x = (x_0, x_1, \ldots): x_n \in X, \ n \geq 0\}$$

with product σ-algebra $\Sigma^{\mathbb{N}} = \sigma(\mathscr{C})$ which is the smallest σ-algebra of subsets of $X^{\mathbb{N}}$ containing the family $\mathscr{C}$ of all finite-dimensional rectangles

$$\mathscr{C} = \{\{x \in X^{\mathbb{N}}: x_0 \in B_0, \ldots, x_n \in B_n\}: B_0, \ldots, B_n \in \Sigma, \ n \geq 0\}. \qquad (2.13)$$

To see this let for each ω the mapping $n \mapsto \xi_n(\omega)$ denotes the sequence $\xi(\omega) = (\xi_n(\omega))_{n \geq 0} \in X^{\mathbb{N}}$. Thus we obtain the mapping $\xi: \Omega \to X^{\mathbb{N}}$ and ξ is measurable if and only if $\xi_n: \Omega \to X$ is measurable for each $n \in \mathbb{N}$. The distribution of the process ξ is a probability measure μ_ξ on $(X^{\mathbb{N}}, \Sigma^{\mathbb{N}})$ and it is uniquely defined through its values on finite-dimensional rectangles. In other words, the finite-dimensional distributions $\mu_{0,1,\ldots,n} = \mu_{(\xi_0,\ldots,\xi_n)}$

$$\mu_{0,1,\ldots,n}(B_0 \times \cdots \times B_n) = \mathbb{P}(\xi_0 \in B_0, \ldots, \xi_n \in B_n), \quad B_0, \ldots, B_n \in \Sigma, \ n \geq 0,$$

uniquely determine the law of the process, since

$$\mu_\xi(B_0 \times \cdots \times B_n \times X^{\mathbb{N}}) = \mathbb{P}(\xi \in B_0 \times \cdots \times B_n \times X^{\mathbb{N}})$$
$$= \mu_{(\xi_0,\ldots,\xi_n)}(B_0 \times \cdots \times B_n)$$

and all finite-dimensional rectangles $B_0 \times \cdots \times B_n \times X^{\mathbb{N}}$ generate $\Sigma^{\mathbb{N}}$. A transition probability and an initial distribution determine the finite-dimensional distributions of a discrete-time Markov process, as stated in the next theorem. Its proof is based on the monotone class theorem and properties of conditional expectations.

Theorem 2.1 *Let μ be a probability measure, P be a transition probability, and $(\xi_n)_{n \geq 0}$ be defined on a probability space $(\Omega, \mathscr{F}, \mathbb{P})$. Then $(\xi_n)_{n \geq 0}$ is a Markov process with transition probability P and initial distribution μ if and only if*

$$\mathbb{P}(\xi_0 \in B_0, \xi_1 \in B_1, \ldots, \xi_n \in B_n)$$
$$= \int_{B_0} \int_{B_1} \cdots \int_{B_{n-1}} P(x_{n-1}, B_n) P(x_{n-2}, dx_{n-1}) \ldots P(x_0, dx_1) \mu(dx_0)$$
$$(2.14)$$

for all sets $B_0, B_1, \ldots, B_n \in \Sigma, \ n \geq 0$.

2.2.2   Random Mapping Representations

Suppose that $(\vartheta_n)_{n \geq 1}$ is a sequence of i.i.d. random variables with values in a metric space Y. Let ξ_0 be an X-valued random variable independent of $(\vartheta_n)_{n \geq 1}$ and with distribution μ. Consider a measurable function $\kappa: X \times Y \to X$ and define a sequence

ξ_n of X-valued random variables by

$$\xi_n = \kappa(\xi_{n-1}, \vartheta_n), \quad n \geq 1. \tag{2.15}$$

For $\mathscr{F}_n = \sigma(\xi_0, \ldots, \xi_n)$ we have $\mathscr{F}_n \subseteq \sigma(\xi_0, \vartheta_1, \ldots, \vartheta_n)$ and ϑ_{n+1} is independent of $\mathscr{F}_n$ and of ϑ_n. Thus, we see that (ξ_n) has the weak Markov property, by (2.1). Hence, ξ is Markov with transition probability of the form

$$P(x, B) = \mathbb{P}(\kappa(x, \vartheta_1) \in B) = \int_Y \mathbf{1}_B(\kappa(x, y)) \nu(dy)$$

and initial distribution μ. A particular example is the *random walk* on $\mathbb{R}^d$ defined by

$$\xi_n = \xi_{n-1} + \vartheta_n, \quad n \geq 1.$$

From Proposition 2.2 it follows that a typical discrete-time Markov process can be defined by a recursive formula as in (2.15).

Theorem 2.2 *Let P be a transition probability on a Borel space (X, Σ). Then there exists a measurable function $\kappa \colon X \times [0, 1] \to X$ such that for any sequence (ϑ_n) of independent random variables with uniform distribution on $[0, 1]$ and for any $x \in X$ the sequence defined by (2.15) is a discrete-time Markov process with transition probability P and initial distribution δ_x.*

To define a discrete-time Markov process with transition probability P on a Borel space (X, Σ), we can take Ω, by Theorem 2.2, to be the countable product $[0, 1]^{\mathbb{N}}$ of the unit interval $[0, 1]$ with the product σ-algebra $\mathscr{F} = \mathscr{B}([0, 1))^{\mathbb{N}}$ and the measure $\mathbb{P}$ as a countable product of the Lebesgue measure on $[0, 1]$. Then we define a sequence of i.i.d. random variables $\vartheta_n \colon \Omega \to [0, 1], n \geq 0$, by $\vartheta_n(\omega) = \omega_n$ for $\omega = (\omega_n)_{n \geq 0} \in \Omega$, and ξ_n by (2.15).

2.2.3 Canonical Processes

We next construct a canonical discrete-time Markov process with a given transition probability and an initial distribution on an arbitrary measurable space (X, Σ). The existence of the process with given finite-dimensional distributions amounts to showing that there is a probability measure on the product space $(X^{\mathbb{N}}, \Sigma^{\mathbb{N}})$ satisfying (2.14) with ξ being the identity mapping. This can be realized through the discrete-time version of the Kolmogorov extension theorem. For its proof we refer the reader to [55]. Note that no regularity condition is needed on the state space.

Theorem 2.3 (Ionescu–Tulcea) *Let P be a transition probability on a measurable space (X, Σ). For every $x \in X$ there exists a unique probability measure $\mathbb{P}_x$ on the product space $(X^{\mathbb{N}}, \Sigma^{\mathbb{N}})$ such that*

$$\mathbb{P}_x(B_0 \times \cdots \times B_n \times X^{\mathbb{N}})$$

$$= 1_{B_0}(x) \int_{B_1} \cdots \int_{B_{n-1}} P(x_{n-1}, B_n) P(x_{n-2}, dx_{n-1}) \ldots P(x, dx_1)$$

for all $B_0, \ldots, B_n \in \Sigma, n \geq 0$. Moreover, for every set $A \in \Sigma^{\mathbb{N}}$, the map $x \mapsto \mathbb{P}_x(A)$ is Σ-measurable.

The function $\mathbb{P}_x(A)$ is a transition probability from (X, Σ) to $(X^{\mathbb{N}}, \Sigma^{\mathbb{N}})$. Thus, by Proposition 2.1, for each probability measure μ on (X, Σ) there exists a unique probability measure $\mathbb{P}_\mu$ on $(X \times X^{\mathbb{N}}, \Sigma \otimes \Sigma^{\mathbb{N}})$ such that

$$\mathbb{P}_\mu(B \times A) = \int_B \mathbb{P}_x(A) \, \mu(dx), \quad B \in \Sigma, \ A \in \Sigma^{\mathbb{N}}.$$

The sequence space

$$\Omega = X^{\mathbb{N}} = \{\omega = (x_0, x_1, \ldots) \colon x_n \in X, \ n \geq 0\}$$

is called a *canonical space* and the coordinate mappings $\xi_n(\omega) = x_n, n \geq 0$, define a process $\xi = (\xi_n)_{n \geq 0}$ on $(\Omega, \mathscr{F}) = (X^{\mathbb{N}}, \Sigma^{\mathbb{N}})$ with distribution $\mathbb{P}_\mu$, called a *Markov canonical process*. Note that if we let the process start at $x \in X$ so that $\mu = \delta_x$ then we have $\mathbb{P}_{\delta_x} = \mathbb{P}_x$ and if $A \in \Sigma^{\mathbb{N}}$ is the rectangle

$$A = \{\omega \colon \xi_0(\omega) \in B_0, \ldots, \xi_n(\omega) \in B_n\}$$

then we have

$$\mathbb{P}_\mu(A) = \mathbb{P}_\mu(\xi_0 \in B_0, \xi_1 \in B_1, \ldots, \xi_n \in B_n)$$

$$= \int_{B_0} \int_{B_1} \cdots \int_{B_{n-1}} P(x_{n-1}, B_n) P(x_{n-2}, dx_{n-1}) \ldots P(x_0, dx_1) \mu(dx_0).$$

Consequently, for each probability measure $\mathbb{P}_\mu$ on $(\Omega, \mathscr{F})$ the process $\xi = (\xi_n)_{n \geq 0}$ is homogeneous Markov with transition probability P and initial distribution μ. We write $\mathbb{E}_x$ for the integration with respect to $\mathbb{P}_x, x \in X$, and we have

$$\mathbb{E}_\mu(\eta) = \int_X \mathbb{E}_x(\eta) \, \mu(dx)$$

for any bounded or non-negative random variable $\eta \colon \Omega \to \mathbb{R}$. In particular, we can rewrite the Markov property as: for each n and any bounded measurable g defined on the sequence space we have

$$\mathbb{E}_\mu(g(\xi_n, \xi_{n+1}, \ldots) | \mathscr{F}_n) = h(\xi_n), \quad \text{where } h(x) = \mathbb{E}_x(g(\xi_0, \xi_1, \ldots)).$$

2.3 Continuous-Time Markov Processes

2.3.1 Basic Definitions

A family $\xi(t)$, $t \in [0, \infty)$, of X-valued random variables is called a *continuous-time stochastic process* with state space X, where (X, Σ) is a measurable space. For each $t \geq 0$, let $\mathscr{F}_t = \sigma(\xi(r) : r \leq t)$ be the σ-algebra generated by all random variables $\xi(r), 0 \leq r \leq t$. Since $\mathscr{F}_s \subseteq \mathscr{F}_t$ for all $0 \leq s < t$, the collection $\mathscr{F}_t, t \geq 0$, is called a *history* of the process or a *filtration*. If X is a Borel space then the process $\xi = \{\xi(t) : t \geq 0\}$ is said to be *right-continuous (càdlàg)* if its sample paths $t \mapsto \xi(t)(\omega)$, $\omega \in \Omega$, are right-continuous (càdlàg).

Let $\xi = \{\xi(t) : t \geq 0\}$ be an X-valued process defined on a probability space $(\Omega, \mathscr{F}, \mathbb{P})$. The process ξ is said to be a *Markov process* if for any times s, t we have

$$\mathbb{P}(\xi(s + t) \in B | \mathscr{F}_s) = \mathbb{P}(\xi(s + t) \in B | \xi(s)), \quad B \in \Sigma. \tag{2.16}$$

By the monotone class theorem the Markov property (2.16) holds if and only if

$$\mathbb{E}(g(\xi(s + t)) | \mathscr{F}_s) = \mathbb{E}(g(\xi(s + t)) | \xi(s))$$

for all $g \in B(X)$, where $B(X)$ is the space of bounded and measurable functions $g : X \to \mathbb{R}$. Note that a process is Markov if and only if for any $0 \leq s_1 < \cdots < s_n$ and $g \in B(X)$

$$\mathbb{E}(g(\xi(s_n + t)) | \xi(s_1), \ldots, \xi(s_n)) = \mathbb{E}(g(\xi(s_n + t)) | \xi(s_n)).$$

A family $P = \{P(t, \cdot) : t \geq 0\}$ of transition probabilities on (X, Σ) is said to be a *transition (probability) function* if it satisfies the *Chapman–Kolmogorov equation*

$$P(s + t, x, B) = \int_X P(s, y, B) P(t, x, dy), \quad s, t \geq 0, \tag{2.17}$$

and $P(0, x, B) = \delta_x(B)$ for every $x \in X$, $B \in \Sigma$. We say that ξ is a *homogeneous Markov process* with transition function P and starting at x if

$$\mathbb{P}(\xi(0) = x) = 1 \quad \text{and} \quad \mathbb{P}(\xi(s + t) \in B | \mathscr{F}_s) = P(t, \xi(s), B) \tag{2.18}$$

for all $B \in \Sigma$ and for all times s and t. Equivalently,

$$\mathbb{E}(g(\xi(s + t)) | \mathscr{F}_s) = \int_X g(y) P(t, \xi(s), dy)$$

for all $g \in B(X)$, $s, t \geq 0$. We interpret $P(t, x, B)$ as the probability that the stochastic process ξ moves from state x at time 0 to a state in B at time t. The transition function P defines a family of bounded linear operators

$$T(t)g(x) = \int_X g(y)P(t, x, dy)$$

on the Banach space $B(X)$ with supremum norm $\| \cdot \|_u$. It follows from the Chapman–Kolmogorov equation (2.17) that $T(t), t \geq 0$, is a *semigroup*, i.e.

$$T(s + t)g = T(t)(T(s)g), \quad g \in B(X), \ s, t \geq 0.$$

The family $\{T(t)\}_{t \geq 0}$ is called the *transition semigroup associated to ξ*.

A random variable $\tau \colon \Omega \to [0, \infty]$ is called a *stopping time* for the filtration $(\mathscr{F}_t)$ if it satisfies $\{\tau \leq t\} \in \mathscr{F}_t$ for all $t \geq 0$. The σ-algebra $\mathscr{F}_\tau$ giving the history known up to time τ is defined as

$$\mathscr{F}_\tau = \{A \in \mathscr{F} \colon A \cap \{\tau \leq t\} \in \mathscr{F}_t \text{ for all } t \geq 0\}.$$

If τ_1 and τ_2 are two stopping times and $\tau_1 \leq \tau_2$ then $\mathscr{F}_{\tau_1} \subseteq \mathscr{F}_{\tau_2}$. A Markov process ξ is said to be *strong Markov* if for any stopping time τ, the *strong Markov property at τ* holds

$$\mathbb{P}(\xi(\tau + t) \in B | \mathscr{F}_\tau) = \mathbb{P}(\xi(\tau + t) \in B | \xi(\tau)) \quad \text{a.s. on } \tau < \infty \tag{2.19}$$

for all $B \in \Sigma$ and all times t. If X is a Borel space and the process has right-continuous sample paths then $\xi(\tau)$ is $\mathscr{F}_\tau$-measurable on $\{\tau < \infty\}$ for any stopping time τ for the filtration $(\mathscr{F}_t)$.

As in the discrete-time case we can consider a canonical space $\Omega = X^{\mathbb{R}_+}$, which now is the space of all functions $\omega \colon \mathbb{R}_+ \to X$, with product σ-algebra $\Sigma^{\mathbb{R}_+} = \sigma(\mathscr{C})$ which is the smallest σ-algebra of subsets of $X^{\mathbb{R}_+}$ containing the family $\mathscr{C}$ of all cylinder sets:

$$\{\omega \in X^{\mathbb{R}_+} \colon \omega(t_0) \in B_0, \ldots, \omega(t_n) \in B_n\}, \quad B_0, \ldots, B_n \in \Sigma, \ 0 \leq t_0 < \cdots < t_n, \ n \geq 0.$$

Let $\mathscr{F}_t = \sigma(\xi(r) : r \leq t), t \geq 0$, where the canonical process $\xi(t), t \geq 0$, is defined as the identity map on $(X^{\mathbb{R}_+}, \Sigma^{\mathbb{R}_+})$. Suppose that P is a transition probability function on a Borel space (X, Σ). It follows from a continuous-time version of the Kolmogorov extension theorem [55] that for every $x \in X$ there exists a probability measure $\mathbb{P}_x$ on $(\Omega, \mathscr{F})$ such that $\xi(t), t \geq 0$, is a homogeneous Markov process with transition function P and starting at x. Moreover, for every set $A \in \Sigma^{\mathbb{N}}$, the map $x \mapsto \mathbb{P}_x(A)$ is Σ-measurable. We have

$$\mathbb{P}_x(\xi(t) \in B) = P(t, x, B), \quad t \geq 0, \ B \in \Sigma,$$

and

$$\mathbb{P}_x(\xi(t_1) \in B_1, \dots, \xi(t_n) \in B_n)$$

$$= \int_{B_1} \cdots \int_{B_{n-1}} P(t_n - t_{n-1}, x_{n-1}, B_n) P(t_{n-1} - t_{n-2}, x_{n-2}, dx_{n-1}) \dots P(t_1, x, dx_1)$$

for all $0 < t_1 < \cdots < t_n$, $B_1, \dots, B_n \in \Sigma$, $n \in \mathbb{N}$.

Given a transition function on a Borel space, for each $x \in X$, there exists a probability space $(\Omega, \mathcal{F}, \mathbb{P}_x)$ and a homogeneous Markov process $\{\xi(t): t \geq 0\}$ defined on $(\Omega, \mathcal{F}, \mathbb{P}_x)$ with transition function P and starting at x. We will also denote the process $\xi(t)$ started at x by $\xi_x(t)$. Then the family of processes $\xi = \{\xi_x(t): t \geq 0, \ x \in X\}$ is called a *Markov family*. We will simply say that $\xi(t)$ is a Markov process with state space X defined on $(\Omega, \mathcal{F}, \mathbb{P}_x)$. However, in general, a transition function is unknown in advance and we need to construct the processes directly from other processes.

2.3.2 Processes with Stationary and Independent Increments

Our first example is a continuous time extension of random walks. An $\mathbb{R}^d$-valued process $\xi(t)$, $t \geq 0$, is said to have *independent increments*, if $\xi(s + t) - \xi(s)$ is independent of $\mathcal{F}_s = \sigma(\xi(r): r \leq s)$ for all $t, s \geq 0$. Given such a process, we have by (2.1)

$$\mathbb{P}(\xi(s + t) \in B | \mathcal{F}_s) = \mathbb{P}(\xi(s + t) - \xi(s) + \xi(s) \in B | \xi(s)) = \mathbb{P}(\xi(s + t) \in B | \xi(s))$$

for $B \in \mathcal{B}(\mathbb{R}^d)$, since $\xi(s + t) = \xi(s + t) - \xi(s) + \xi(s)$ and $\xi(s + t) - \xi(s)$ is independent of $\mathcal{F}_s$, which shows that ξ has the Markov property.

The process ξ has *stationary increments*, if the distribution of $\xi(s + t) - \xi(s)$ is the same as the distribution of $\xi(t) - \xi(0)$ for all $s, t \geq 0$. In that case

$$\mathbb{E}\mathbf{1}_B(\xi(s + t) - \xi(s) + x) = \mathbb{E}\mathbf{1}_B(\xi(t) - \xi(0) + x)$$

for all x and s, t. In particular, we have $\xi(s + t) - \xi(0) = \xi(s + t) - \xi(t) + \xi(t) - \xi(0)$ and the random variables $\xi(s + t) - \xi(t)$ and $\xi(t) - \xi(0)$ are independent. Hence, if μ_t is the distribution of $\xi(t) - \xi(0)$, then μ_{s+t} is the convolution of μ_s and μ_t, i.e.

$$\mu_{s+t}(B) = (\mu_s * \mu_t)(B) = \int \mu_s(B - x)\mu_t(dx),$$

and

$$P(t, x, B) = \int_{\mathbb{R}^d} \mathbf{1}_B(x + y)\mu_t(dy).$$

A process is called a *Lévy process* if it has stationary independent increments, it starts at zero, i.e. $\xi(0) = 0$ a.s., and it is *continuous in probability*, i.e. for every

$\varepsilon > 0$ we have

$$\lim_{t \to 0} \mathbb{P}(|\xi(t)| > \varepsilon) = 0.$$

2.3.3 Markov Jump-Type Processes

In this section, we provide a simple construction of pure jump-type processes with bounded jump rate function. They were introduced in Sect. 1.3 and are particular examples of PDMPs defined in Sect. 1.19. Here we show that they are Markov processes. Suppose that (X, Σ) is a measurable space, P is a transition probability on X and that φ is a bounded non-negative measurable function. Set $\lambda = \sup\{\varphi(x) \colon x \in X\}$ and define the transition probability $\bar{P}$ by

$$\bar{P}(x, B) = \lambda^{-1}(\varphi(x)P(x, B) + (\lambda - \varphi(x))\delta_x(B)), \quad x \in X, B \in \Sigma. \quad (2.20)$$

Let $(\xi_n)_{n \geq 0}$ be a discrete-time Markov process with transition probability $\bar{P}$ and $(\sigma_n)_{n \geq 1}$ be a sequence of independent random variables, exponentially distributed with mean λ^{-1}, and independent of the sequence $(\xi_n)_{n \geq 0}$. We set $\tau_0 = 0$ and we define

$$\xi(t) = \xi_{n-1} \text{ if } \tau_{n-1} \leq t < \tau_n, \quad \text{where } \tau_n = \sum_{k=1}^{n} \sigma_k, \quad n \geq 1.$$

It follows from the strong law of large numbers that $\tau_n \to \infty$, as $n \to \infty$, a.s. The sample paths of the process ξ are constant between consecutive τ_n and the random variables σ_n are called holding times.

Let $N(t)$ be the number of jump times τ_n in the time interval $(0, t]$, i.e. we have

$$N(t) = \max\{n \geq 0 \colon \tau_n \leq t\} = \sum_{n=0}^{\infty} \mathbf{1}_{(0,t]}(\tau_n).$$

Then $N(t) = 0$ if $t < \tau_1$, and

$$N(t) = n \quad \Longleftrightarrow \quad \tau_n \leq t < \tau_{n+1}.$$

We show that $N(t)$ is Poisson distributed with parameter λt, i.e.

$$\mathbb{P}(N(t) = n) = e^{-\lambda t}\frac{(\lambda t)^n}{n!}, \quad n \geq 0.$$

We have $\mathbb{P}(N(t) = 0) = \mathbb{P}(\tau_1 > t) = e^{-\lambda t}$ and for $n \geq 1$

$$\mathbb{P}(N(t) = n) = \mathbb{P}(\tau_n \leq t < \tau_{n+1}) = \mathbb{P}(\tau_{n+1} > t) - \mathbb{P}(\tau_n > t),$$

since $\{\tau_n > t\} \subseteq \{\tau_{n+1} > t\}$. The random variable τ_n, being the sum of n independent exponentially distributed random variables with parameter λ, has a gamma distribution with density

$$f_{\tau_n}(x) = \frac{\lambda^n x^{n-1}}{(n-1)!} e^{-\lambda x} \quad \text{for } x \geq 0.$$

Hence,

$$\mathbb{P}(\tau_n > t) = \int_t^\infty \frac{\lambda^n x^{n-1}}{(n-1)!} e^{-\lambda x} dx = -e^{-\lambda t} \frac{(\lambda t)^n}{n!} + \mathbb{P}(\tau_{n+1} > t).$$

We can write $\xi(t) = \xi_{N(t)}$ for $t \geq 0$. In particular, if we take $X = \mathbb{N}$ and the trivial Markov chain $\xi_n = n$ for all n, then we have $\xi(t) = N(t)$ for all $t \geq 0$ and $N(t)$, $t \geq 0$, is a Poisson process with intensity $\lambda > 0$. If $(\xi_n)_{n \geq 0}$ is a random walk then $\xi(t)$ is the compound Poisson process as in (1.3). The Poisson process N has stationary independent increments, equivalently, for any n, t, s and $A \in \sigma(N(r): r \leq s)$

$$\mathbb{P}(\{N(t+s) - N(s) = n\} \cap A) = \mathbb{P}(N(t) = n)\mathbb{P}(A).$$

Consider again the general process $\xi(t) = \xi_{N(t)}$ for $t \geq 0$. We denote by $\xi_x(t)$ the particular process $\xi(t)$ starting at $\xi(0) = \xi_0 = x$. We can easily calculate the distribution of $\xi_x(t)$. By independence of $N(t)$ and (ξ_n), we obtain

$$\mathbb{P}(\xi_x(t) \in B) = \mathbb{P}(\xi(t) \in B | \xi_0 = x) = \sum_{n=0}^\infty \mathbb{P}(N(t) = n)\mathbb{P}(\xi_n \in B | \xi_0 = x).$$

Since the random variable $N(t)$ is Poisson distributed with parameter λt, we have

$$P(t, x, B) = \sum_{n=0}^\infty e^{-\lambda t} \frac{(\lambda t)^n}{n!} \bar{P}^n(x, B), \tag{2.21}$$

where $\bar{P}^n$ is the nth step transition probability.

We now check that condition (2.18) holds with $\mathscr{F}_s = \sigma(\xi(r): r \leq s)$. We may write

$$\mathbb{E}(\mathbf{1}_B(\xi(t+s))|\mathscr{F}_s) = \mathbb{E}(\mathbf{1}_B(\xi_{N(t+s)})|\mathscr{F}_s) = \mathbb{E}(\mathbf{1}_B(\xi_{N(t+s)-N(s)+N(s)})|\mathscr{F}_s),$$

which gives

$$\mathbb{E}(\mathbf{1}_B(\xi(t+s))|\mathscr{F}_s) = \sum_{n=0}^\infty \mathbb{E}(\mathbf{1}_{\{N(t+s)-N(s)=n\}} \mathbf{1}_B(\xi_{n+N(s)})|\mathscr{F}_s).$$

Since $N(t+s) - N(s)$ is independent of $\sigma(N(r): r \le s)$ and $\sigma(\xi_n: n \ge 0)$, it is also independent of $\mathscr{F}_s$. It is Poisson distributed with parameter λt. Thus

$$\mathbb{E}(\mathbf{1}_B(\xi(t+s))|\mathscr{F}_s) = \sum_{n=0}^{\infty} \mathbb{P}(N(t+s) - N(s) = n)\mathbb{E}(\mathbf{1}_B(\xi_{n+N(s)})|\mathscr{F}_s)$$

$$= \sum_{n=0}^{\infty} e^{-\lambda t} \frac{(\lambda t)^n}{n!} \mathbb{E}(\mathbf{1}_B(\xi_{n+N(s)})|\mathscr{F}_s).$$

The family

$$\{A_1 \cap A_2 \cap \{N(s) = m\}: A_1 \in \sigma(N(r) \le s), \quad A_2 \in \sigma(\xi_k: k \le m), \quad m \ge 0\}$$

generates the σ-algebra $\mathscr{F}_s$. Thus, it is enough to show that

$$\int_{A_1 \cap A_2 \cap \{N(s)=m\}} \mathbf{1}_B(\xi_{n+N(s)}) \, d\mathbb{P} = \int_{A_1 \cap A_2 \cap \{N(s)=m\}} \bar{P}^n(\xi(s), B) \, d\mathbb{P}.$$

To this end, observe that $\xi(s) = \xi_m$ on $\{N(s) = m\}$ and $A_1 \cap \{N(s) = m\}$ is independent of A_2 and ξ_m. This together with the Markov property for $(\xi_n)_{n \ge 0}$ leads to

$$\int_{A_1 \cap A_2 \cap \{N(s)=m\}} \bar{P}^n(\xi(s), B) \, d\mathbb{P} = \mathbb{P}(A_1 \cap \{N(s) = m\}) \int_{A_2} \bar{P}^n(\xi_m, B) \, d\mathbb{P}$$

$$= \mathbb{P}(A_1 \cap \{N(s) = m\}) \int_{A_2} \mathbf{1}_B(\xi_{n+m}) \, d\mathbb{P}.$$

Now, making use of the independence of $A_1 \cap \{N(s) = m\}$ and $A_2 \cap \{\xi_{n+m} \in B\}$, completes the proof of the Markov property.

2.3.4 Generators and Martingales

Assume that $\xi(t)$ is a Markov process with state space X defined on $(\Omega, \mathscr{F}, \mathbb{P}_x)$. Let

$$T(t)g(x) = \mathbb{E}_x g(\xi(t)) = \int_X g(y) P(t, x, dy), \quad g \in B(X), \ x \in X, \ t \ge 0,$$

where $\mathbb{E}_x$ is the expectation with respect to $\mathbb{P}_x$. Consider the class $\mathscr{D}(L)$ of all bounded and measurable functions $g: X \to \mathbb{R}$ such that the limit

$$\lim_{t \downarrow 0} \frac{\mathbb{E}_x(g(\xi(t))) - g(x)}{t}$$

exists uniformly for all $x \in X$. We denote the limit by $Lg(x)$ and we call L the *infinitesimal generator* of the Markov process ξ. In particular,

$$Lg(x) = \lim_{t \downarrow 0} \frac{1}{t} \int_X (g(y) - g(x)) P(t, x, dy)$$

for all $x \in X$, whenever $g \in \mathscr{D}(L)$.

For a pure jump-type process with bounded jump rate function φ as considered in Sect. 2.3.3 and with transition function of the form (2.21) we have

$$\int_X (g(y) - g(x)) P(t, x, dy) = (e^{-\lambda t} - 1) g(x) + e^{-\lambda t} \lambda t \int_X g(y) \bar{P}(x, dy)$$

$$+ \sum_{n=2}^{\infty} e^{-\lambda t} \frac{(\lambda t)^n}{n!} \int_X g(y) \bar{P}^n(x, dy),$$

which shows that $\mathscr{D}(L)$ consists of all bounded measurable functions and that

$$Lg(x) = \lambda \int_X (g(y) - g(x)) \bar{P}(x, dy), \quad x \in X, \ g \in B(X).$$

Using (2.20) we conclude that the infinitesimal generator of this process is of the form

$$Lg(x) = \varphi(x) \int_X (g(y) - g(x)) P(x, dy), \quad x \in X, \ g \in B(X).$$

The significance of the infinitesimal generator $(L, \mathscr{D}(L))$ is related to the *Dynkin formula*, which states that if $g \in \mathscr{D}(L)$ then the process

$$\eta(t) = g(\xi(t)) - g(\xi(0)) - \int_0^t Lg(\xi(r)) \, dr$$

is a *martingale*, i.e. the random variable $\eta(t)$ is integrable, $\mathscr{F}_t$-measurable for each $t \geq 0$, and

$$\mathbb{E}(\eta(t + s) | \mathscr{F}_s) = \eta(s)$$

for all $t, s \geq 0$. In particular, the Dynkin formula holds if $\xi(t)$ is a Markov process with a metric state space and right-continuous paths. To see this take g and Lg bounded and observe that

$$\mathbb{E}_x(\eta(t + s) | \mathscr{F}_s) = \mathbb{E}_x(g(\xi(t + s)) | \mathscr{F}_s) - g(x) - \int_0^{t+s} \mathbb{E}_x(Lg(\xi(r)) | \mathscr{F}_s) \, dr.$$

Since $Lg(\xi(r))$ is $\mathscr{F}_s$-measurable for all $r \leq s$, we can write

$$\int_0^{t+s} \mathbb{E}_x(Lg(\xi(r))|\mathscr{F}_s)\,dr = \int_0^s Lg(\xi(r))\,dr + \int_0^t \mathbb{E}_x(Lg(\xi(s+r))|\mathscr{F}_s)\,dr,$$

and, by the Markov property, we have

$$\mathbb{E}_x(\eta(t+s)|\mathscr{F}_s) = T(t)g(\xi(s)) - g(x) - \int_0^s Lg(\xi(r))\,dr - \int_0^t T(r)(Lg)(\xi(s))\,dr,$$

which gives the claim, by using the identity (see (3.4) in Sect. 3.1.2)

$$T(t)g = g + \int_0^t T(r)(Lg)\,dr, \quad t \geq 0.$$

There are several different versions of generators for Markov processes. For example, one can consider instead of the uniform convergence in $B(X)$, the pointwise convergence or the so-called bounded pointwise convergence (see [35]). Another approach, given by [28], introduces the extended generator using the concept of local martingales and allowing unbounded functions in the domain of the generator; here we adopt this definition. Let $M(X)$ be the space of all measurable functions $g: X \to \mathbb{R}$. An operator $\widetilde{L}$ is called the *extended generator* of the Markov process ξ, if its domain $\mathscr{D}(\widetilde{L})$ consists of those $g \in M(X)$ for which there exists $f \in M(X)$ such that for each $x \in X, t > 0$,

$$\mathbb{E}_x(g(\xi(t))) = g(x) + \mathbb{E}_x\left(\int_0^t f(\xi(r))\,dr\right)$$

and

$$\int_0^t \mathbb{E}_x(|f(\xi(r))|)\,dr < \infty,$$

in which case we define $\widetilde{L}g = f$.

2.3.5   Existence of PDMPs

In this section, we consider the general setting from Sect. 1.19. We assume that (X, Σ) is a Borel space and that (π, Φ, P) are three characteristics representing, respectively, a semiflow, a survival function, and a jump distribution, being a transition probability from $X \cup \Gamma$ to X where Γ is the active boundary. We assume that $P(x, X \setminus \{x\}) = 1$ for all $x \in X \cup \Gamma$. Since (X, Σ) is a Borel space, we can find a measurable mapping $\kappa: (X \cup \Gamma) \times [0, 1] \to X$ such that

$$P(x, B) = \text{Leb}\{r \in [0, 1]: \kappa(x, r) \in B\}, \quad x \in X \cup \Gamma, \ B \in \Sigma. \tag{2.22}$$

We extend the state space and the characteristics to $(X_\Delta, \Sigma_\Delta)$ as described in Sect. 1.19. We define $\kappa(\Delta, r) = \Delta$ for $r \in [0, 1]$. Thus formula (2.22) remains valid for $x \in X_\Delta \cup \Gamma$. We extend every function g defined on X to X_Δ by setting $g(\Delta) = 0$. For each $x \in X_\Delta$ we define the generalized inverse of $t \mapsto \Phi_x(t)$ by

$$\Phi_x^\leftarrow(q) = \inf\{t : \Phi_x(t) \le q\}, \quad q \ge 0. \tag{2.23}$$

If ϑ is a random variable uniformly distributed on $[0, 1]$, then we have

$$\mathbb{P}(\sigma > t) = \Phi_x(t), \quad t \in [0, \infty], \quad \text{where } \sigma = \Phi_x^\leftarrow(\vartheta).$$

Let $(\Omega, \mathscr{F}, \mathbb{P})$ be a probability space and let $(\vartheta_n)_{n \ge 1}$ be a sequence of independent random variables with uniform distribution on $[0, 1]$. We define $\tau_0 = \sigma_0 = 0, \xi_0 = x$ and

$$\sigma_1 = \Phi_{\xi_0}^\leftarrow(\vartheta_1), \quad \tau_1 = \sigma_1 + \tau_0,$$

and we set $\xi(t) = \pi(t - \tau_0, \xi_0)$ for $t < \tau_1$. Since the function $t \mapsto \pi(t, x)$ has a left-hand limit, which belongs to the set $X_\Delta \cup \Gamma$, we can define

$$\xi_1 = \kappa(\xi(\tau_1^-), \vartheta_2) \quad \text{and} \quad \xi(\tau_1) = \xi_1.$$

On the set $\{\tau_1 = \infty\}$ the process is defined for all times t. On $\{\tau_1 < \infty\}$ we continue the construction of the process inductively. We define $\sigma_2 = \Phi_{\xi_1}^\leftarrow(\vartheta_3), \tau_2 = \sigma_2 + \tau_1$, and we set

$$\xi(t) = \pi(t - \tau_1, \xi_1) \text{ if } \tau_1 \le t < \tau_2, \quad \xi(\tau_2) = \xi_2 = \kappa(\xi(\tau_2^-), \vartheta_4),$$

and so on. Consequently, we define the minimal process $\{\xi(t)\}_{t \ge 0}$ starting at $\xi(0) = x$ by

$$\xi(t) = \begin{cases} \pi(t - \tau_n, \xi_n), & \text{if } \tau_n \le t < \tau_{n+1}, \ n \ge 0, \\ \Delta, & \text{if } t \ge \tau_\infty, \end{cases} \tag{2.24}$$

where

$$\tau_n = \sigma_n + \tau_{n-1}, \quad \sigma_n = \Phi_{\xi_{n-1}}^\leftarrow(\vartheta_{2n-1}), \quad \xi_n = \kappa(\xi(\tau_n^-), \vartheta_{2n}), \quad n \ge 1. \tag{2.25}$$

Let $N(t)$ be the number of jump times τ_n in the time interval $[0, t]$

$$N(t) = \sup\{n \ge 0 : \tau_n \le t\}. \tag{2.26}$$

Then $N(t) = 0$ if $t < \tau_1$, $N(t) = n$ if and only if $\tau_n \le t < \tau_{n+1}$, and $N(t) = \infty$ for $t \ge \tau_\infty$. If we set $\xi_\infty = \Delta$ and $\tau_{\infty+1} = \infty$ then we have

$$\xi(t) = \pi(t - \tau_n, \xi_n) \quad \text{on} \quad \{\tau_n \le t < \tau_{n+1}\}$$

for some $n \in \bar{\mathbb{N}} = \{0, 1, \ldots\} \cup \{\infty\}$, $t \in [0, \infty]$, which we can rewrite as

$$\xi(t) = \pi(t - \tau_{N(t)}, \xi_{N(t)}), \quad t \in [0, \infty].$$

Theorem 2.4 *The minimal process $\xi(t)$, $t \geq 0$, as defined in (2.24) is a strong Markov process.*

We outline the main steps of the proof. It is similar to the proof given in [28, 29] for processes with Euclidean state space. Let $\mathscr{F}_t$ be the filtration generated by $\xi(t)$. Then each τ_n is a stopping time. Observe that for any $t \geq 0$

$$\mathscr{F}_t = \sigma(1_B(\xi_k)1_{[0,r]}(\tau_k): k \in \bar{\mathbb{N}}, \ B \in \Sigma_\Delta, \ 0 \leq r \leq t).$$

It is easy to see that $\mathscr{F}_{\tau_n} = \sigma(\tau_k, \xi_k: k \leq n)$, $n \in \bar{\mathbb{N}}$, and that

$$\mathscr{F}_s \cap \{\tau_n \leq s < \tau_{n+1}\} = \mathscr{F}_{\tau_n} \cap \{\tau_n \leq s < \tau_{n+1}\}, \quad n \in \bar{\mathbb{N}}.$$

Note that for $q, r \geq 0$ we have

$$\Phi_x^\leftarrow(q) > r \quad \Longleftrightarrow \quad \Phi_x(r) > q.$$

Thus we obtain

$$\mathbb{P}(\tau_{n+1} > r|\mathscr{F}_{\tau_n}) = \mathbb{P}(\sigma_{n+1} > r - \tau_n|\mathscr{F}_{\tau_n}) = \mathbb{P}(\Phi_{\xi_n}^\leftarrow(\vartheta_{2n+1}) > r - \tau_n|\mathscr{F}_{\tau_n})$$
$$= \Phi_{\xi_n}(r - \tau_n)1_{\{\tau_n \leq r\}} + 1_{\{\tau_n > r\}},$$

which implies that

$$\mathbb{P}(\tau_{n+1} > t + s|\mathscr{F}_s) = \Phi_{\xi(s)}(t) \quad \text{on } \{\tau_n \leq s < \tau_{n+1}\} \tag{2.27}$$

and leads to the weak Markov property. To show the strong Markov property, we take a stopping time τ. We have

$$\mathscr{F}_\tau \cap \{\tau_n \leq \tau < \tau_{n+1}\} = \mathscr{F}_{\tau_n} \cap \{\tau_n \leq \tau < \tau_{n+1}\}, \quad n \in \bar{\mathbb{N}},$$

and for each n there exists $\mathscr{F}_{\tau_n}$-measurable random variable ζ_n such that

$$\tau 1_{\{\tau < \tau_{n+1}\}} = \zeta_n 1_{\{\tau < \tau_{n+1}\}},$$

so that (2.27) remains valid for $s = \tau$.

Let $\mathbb{P}_x$ be the distribution of the process $\xi(t)$ starting at x. The transition probability function is given by

$$P(t, x, B) = \mathbb{P}_x(\xi(t) \in B) = \mathbb{P}_x(\xi(t) \in B, \ t < \tau_\infty) + \mathbb{P}_x(\xi(t) \in B, \ t \geq \tau_\infty).$$

For $x = \Delta$ we have $\xi(t) = \Delta$, thus $P(t, \Delta, B) = \delta_\Delta(B)$ for all $t \geq 0$. If $x \in X$ and $\Delta \notin B$, then $\mathbb{P}_x(\xi(t) \in B, \, t \geq \tau_\infty) = 0$ for all t. Thus for any $x \in X$ and $B \in \Sigma$ we have

$$P(t, x, B) = \mathbb{P}_x(\xi(t) \in B, \, t < \tau_\infty) = \sum_{n=0}^{\infty} \mathbb{P}_x(\xi(t) \in B, \, \tau_n \leq t < \tau_{n+1}). \quad (2.28)$$

Note that if

$$\mathbb{E}_x(N(t)) = \mathbb{E}_x\left(\sum_n \mathbf{1}_{(0,t]}(\tau_n)\right) < \infty \quad \text{for all } t > 0, \, x \in X, \quad (2.29)$$

then the process ξ is non-explosive, i.e. $\tau_\infty = \infty$ a.s. In that case we have $P(t, x, X) = 1$ for all $t > 0$ and $x \in X$.

Remark 2.2 If $P(x, \{x\}) \neq 0$ for some $x \in X$ then we can extend the state space X to $\widehat{X} = X \times \{0, 1\}$ and define a transition probability $\widehat{P}$ by

$$\widehat{P}((x, i), B \times \{1 - i\}) = P(x, B), \quad \widehat{P}((x, i), B \times \{i\}) = 0, \quad (x, i) \in (X \cup \Gamma) \times \{0, 1\}.$$

It follows from (2.22) that $\widehat{\kappa} \colon \widehat{X} \times [0, 1] \to \widehat{X}$ given by $\widehat{\kappa}((x, i), r) = (\kappa(x, r), 1 - i)$ satisfies

$$\widehat{P}((x, i), B \times \{j\}) = \text{Leb}\{r \in [0, 1] \colon \widehat{\kappa}((x, i), r) \in B \times \{j\}\}).$$

We define the semiflow $\widehat{\pi}$ and the survival function $\widehat{\Phi}$ by

$$\widehat{\pi}(t, x, i) = (\pi(t, x), i) \quad \text{and} \quad \widehat{\Phi}_{(x,i)}(t) = \Phi_x(t), \quad (x, i) \in \widehat{X}, \, t \geq 0.$$

Using the characteristics $(\widehat{\pi}, \widehat{\Phi}, \widehat{P})$ we construct the process $\widehat{\xi}(t) = (\xi(t), i(t))$, $t \geq 0$, on $\widehat{X}_\Delta = X_\Delta \times \{0, 1\}$ as in (2.24). It is strong Markov by Theorem 2.4. Its restriction ξ to the state space X remains a strong Markov process.

2.3.6 Transition Functions and Generators of PDMPs

We consider the minimal PDMP ξ with characteristics (π, Φ, P) and jump times (τ_n) as given in Sect. 2.3.5. Let $\mathbb{P}_x$ be the distribution of the process $\xi(t)$ starting at x. For any non-negative measurable functions h defined on $X_\Delta \times [0, \infty]$, we have

$$\mathbb{E}_x[h(\xi(\tau_1), \tau_1)] = \int_{X_\Delta \times [0,\infty]} h(y, s) P(\pi(s^-, x), dy) \Phi_x(ds).$$

We define the transition kernel

$$K(x, B \times J) = \mathbb{E}_x[\mathbf{1}_B(\xi(\tau_1))\mathbf{1}_J(\tau_1)], \quad x \in X_\Delta,$$

for $B \in \Sigma_\Delta$, $J \in \mathcal{B}([0, \infty])$. The strong Markov property of the process $\xi(t)$ at τ_n implies that the sequence $(\xi(\tau_n), \tau_n), n \geq 0$, is a Markov chain on $X_\Delta \times [0, \infty]$ such that for $B \in \Sigma_\Delta$ and $J \in \mathcal{B}([0, \infty])$

$$\mathbb{P}(\xi(\tau_{n+1}) \in B, \tau_{n+1} - \tau_n \in J | \mathscr{F}_{\tau_n}) = K(\xi(\tau_n), B \times J).$$

We have the iterative formula

$$K^n(x, B \times J) = \mathbb{P}_x(\xi(\tau_n) \in B, \tau_n \in J) = \int_{X_\Delta \times [0,\infty]} K^{n-1}(y, B \times (J - s))K(x, dy, ds)$$

for $n \geq 1$, $K^1 = K$, and $K^0(x, B \times J) = \mathbf{1}_B(x)\delta_0(J)$. Note that

$$\mathbb{P}_x(\xi(t) \in B, t < \tau_1) = \mathbf{1}_B(\pi(t, x))\mathbb{P}_x(\tau_1 > t) = \mathbf{1}_B(\pi(t, x))\Phi_x(t).$$

Since $\xi(t) = \pi(t - \tau_n, \xi(\tau_n))$ on $\{\tau_n \leq t < \tau_{n+1}\}$, it follows form (2.28) that

$$P(t, x, B) = \sum_{n=0}^{\infty} \int_{X \times [0,t]} \mathbf{1}_B(\pi(t - s, y))\Phi_y(t - s)K^n(x, dy, ds) \qquad (2.30)$$

for all $x \in X, B \in \Sigma, t > 0$.

We now show that the transition function P of the process ξ satisfies the following *Kolmogorov equation*

$$P(t, x, B) = \mathbf{1}_B(\pi(t, x))\Phi_x(t) + \int_0^t \int_X P(t - s, y, B)K(x, dy, ds) \qquad (2.31)$$

for $x \in X, t > 0, B \in \Sigma$. To this end define for each $n \geq 0$ and $t \geq 0$

$$P_n(t, x, B) = \mathbb{P}_x(\xi(t) \in B, \ t < \tau_{n+1}), \quad x \in X, \ B \in \Sigma. \qquad (2.32)$$

It follows from the monotone convergence theorem and (2.28) that

$$P_n(t, x, B) = \mathbb{E}_x(\mathbf{1}_B(\xi(t))\mathbf{1}_{\{t < \tau_{n+1}\}}) \uparrow \mathbb{E}_x(\mathbf{1}_B(\xi(t))\mathbf{1}_{\{t < \tau_\infty\}}) = P(t, x, B).$$

For any non-negative measurable function g we have

$$\mathbb{E}_x g(\xi(t))\mathbf{1}_{\{t < \tau_{n+1}\}} = \mathbb{E}_x g(\xi(t))\mathbf{1}_{\{t < \tau_1\}} + \mathbb{E}_x g(\xi(t))\mathbf{1}_{\{\tau_1 \leq t < \tau_{n+1}\}}$$

and the strong Markov property implies that

$$\mathbb{E}_x \mathbf{1}_B(\xi(t))\mathbf{1}_{\{\tau_1 \leq t < \tau_{n+1}\}} = \int_{X \times [0,t]} P_{n-1}(t - s, y, B)K(x, dy, ds).$$

Hence, the monotone convergence theorem completes the proof of (2.31).

Let $M(X)_+$ (respectively $B(X)_+$) be the space of all non-negative (bounded) measurable functions on X. We define

$$T_n(t)g(x) = \int_X g(y)P_n(t, x, dy) \tag{2.33}$$

for $t \geq 0$, $x \in X$, $g \in M(X)_+$, $n \geq 0$, where P_n is as in (2.32). Let T be the transition operator corresponding to the jump distribution P. It is defined for $g \in M(X)_+$ by

$$Tg(x) = \int_X g(y)P(x, dy), \quad x \in X.$$

Then, for $t \geq 0$, we have

$$\int_{X\times[0,t]} T_{n-1}(t-s)g(y)K(x, dy, ds) = \int_0^t \int_X T_{n-1}(t-s)g(y)P(\pi(s^-, x), dy)\Phi_x(ds)$$

$$= \int_0^t T(T_{n-1}(t-s)g)(\pi(s^-, x))\Phi_x(ds).$$

We now suppose that the semiflow is continuous in X, i.e. $\pi(s^-, x) = \pi(s, x)$ for all $s < t_*(x)$, $x \in X$. We consider Φ as in (1.38) defined with the help of a jump rate function φ. Then

$$\int_0^t T(T_{n-1}(t-s)g)(\pi(s, x))\Phi_x(ds) = \int_0^t T(T_{n-1}(t-s)g)(\pi(s, x))\varphi(\pi(s, x))\Phi_x(s)\, ds$$

and we obtain

$$T_n(t)\mathbf{1}_B(x) = T_0(t)\mathbf{1}_B(x) + \int_0^t T_0(s)(\varphi T(T_{n-1}(t-s)\mathbf{1}_B))(x)\, ds \tag{2.34}$$

for all $x \in X$, $t \geq 0$, and $n \geq 1$.

We conclude this section with a description of the extended generator $(\widetilde{L}, \mathscr{D}(\widetilde{L}))$ in the case when the active boundary might be non-empty, the survival function is as in (1.41), and the minimal process ξ satisfies (2.29). In particular, if the jump rate function φ in condition (1.41) is bounded then (2.29) holds. Given the active boundary Γ defined in (1.40) we write that $g \in M_\Gamma(X)$ if $g: X \to \mathbb{R}$ is measurable and the function $t \mapsto g(\pi(t, x))$ has a finite limit as $t \to t_*(x)$ for $x \in X$ with finite $t_*(x)$, where $t_*(x)$ is the exit time from X as defined in (1.39). If $g \in M_\Gamma(X)$ has the following properties

(1) for each $x \in X$ the function $t \mapsto g(\pi(t, x))$ is absolutely continuous on $(0, t_*(x))$,
(2) for each $x \in \Gamma$ we have

$$g(x) = \int_X g(y)P(x, dy),$$

(3) for each $t \geq 0$, $x \in X$,

$$\mathbb{E}_x \left(\sum_{\tau_n \leq t} |g(\xi(\tau_n)) - g(\xi(\tau_n^-))| \right) < \infty,$$

then $g \in \mathscr{D}(\widetilde{L})$ and

$$\widetilde{L}g(x) = \widetilde{L}_0 g(x) + \varphi(x) \int_X (g(y) - g(x)) P(x, dy), \quad x \in X, \qquad (2.35)$$

with $\widetilde{L}_0 g$ defined by

$$g(\pi(t, x)) - g(x) = \int_0^t \widetilde{L}_0 g(\pi(s, x)) \, ds, \quad t < t_*(x), \ x \in X.$$

A more general condition instead of (3) characterizes all elements of the domain of the extended generator as defined and showed in [28, 29]. Note that if g is bounded and condition (2.29) holds, then g satisfies condition (3).

Chapter 3
Operator Semigroups

Semigroups of linear operators provide the primary tools in the study of continuous-time Markov processes. They arise as the solutions of the initial value problem for the differential equation $u'(t) = Au(t)$, where A is a linear operator acting on a Banach space. We describe what is generally regarded as the basic theory. We provide basic definitions, examples and theorems characterizing the operators as being the generators of semigroups. The aim here is to provide necessary foundations for studying semigroups on L^1 spaces in the next chapter.

3.1 Generators and Semigroups

3.1.1 Essentials of Banach Spaces and Operators

In this chapter, we assume that $(\mathscr{X}, \|\cdot\|)$ is a real Banach space. That is, $\mathscr{X}$ is a real vector space and $\|\cdot\|$, called the norm, is a non-negative function defined on $\mathscr{X}$ satisfying:

$$\|f\| = 0 \text{ if and only if } f = 0,$$
$$\|cf\| = |c|\,\|f\|, \text{ whenever } c \in \mathbb{R} \text{ and } f \in \mathscr{X},$$
$$\|f + g\| \leq \|f\| + \|g\|, \text{ whenever } f, g \in \mathscr{X},$$

and the metric space $(\mathscr{X}, \rho)$ with $\rho(f, g) = \|f - g\|$ is complete. Usually, $\mathscr{X}$ will be a space of real-valued functions defined on a state space X.

A linear operator A on $\mathscr{X}$ is a linear mapping $A\colon \mathscr{D}(A) \to \mathscr{X}$, where $\mathscr{D}(A)$ is a linear subspace of $\mathscr{X}$, called the domain of A. It is said to be *bounded* if $\mathscr{D}(A) = \mathscr{X}$ and $\|A\| = \sup_{\|f\| \leq 1} \|Af\|$, called the norm of A, is finite. Note that a

© The Author(s) 2017
R. Rudnicki and M. Tyran-Kamińska, *Piecewise Deterministic Processes in Biological Models*, SpringerBriefs in Mathematical Methods,
DOI 10.1007/978-3-319-61295-9_3

linear operator A with $\mathcal{D}(A) = \mathcal{X}$ is bounded if and only if it is *continuous*, i.e. the mapping $f \mapsto Af$ is continuous for all $f \in \mathcal{X}$. The operator A is a *contraction* if $\|A\| \leq 1$. The operator A is said to be *densely defined* if its domain $\mathcal{D}(A)$ is dense in $\mathcal{X}$, so that every $f \in \mathcal{X}$ is a limit of a sequence of elements from $\mathcal{D}(A)$. The operator $(A, \mathcal{D}(A))$ is *closed* if its graph $\{(f, Af): f \in \mathcal{D}(A)\}$ is a closed set in the product space $\mathcal{X} \times \mathcal{X}$ or equivalently if $f_n \in \mathcal{D}(A)$, $n \geq 1$,

$$\lim_{n \to \infty} f_n = f, \quad \text{and} \quad \lim_{n \to \infty} Af_n = g,$$

then $f \in \mathcal{D}(A)$ and $g = Af$.

A linear operator $(A, \mathcal{D}(A))$ on a Banach space $\mathcal{X}$ is said to be *invertible* if there is a bounded operator A^{-1} on $\mathcal{X}$ such that $A^{-1}Af = f$ for all $f \in \mathcal{D}(A)$ and $A^{-1}g \in \mathcal{D}(A)$ and $AA^{-1}g = g$ for all $g \in \mathcal{X}$. Observe that A is invertible if and only if A is closed, A is onto $\{Af: f \in \mathcal{D}(A)\} = \mathcal{X}$ and A is *one to one*, i.e. if $Af = 0$ then $f = 0$. The set $\{Af: f \in \mathcal{D}(A)\}$, called the *range of* $(A, \mathcal{D}(A))$, is denoted by $\mathcal{R}(A)$, and the set $\{f \in \mathcal{D}(A): Af = 0\}$, called the *nullspace*, by $\mathcal{N}(A)$. Analogous definitions are valid for operators acting between two different Banach spaces.

We denote by $\mathcal{X}^*$ the space of all continuous linear functionals $\alpha: \mathcal{X} \to \mathbb{R}$. It is a real Banach space with the norm

$$\|\alpha\| = \sup_{\|f\| \leq 1} |\alpha(f)|, \quad \alpha \in \mathcal{X}^*,$$

and it is called the *dual space* of $\mathcal{X}$. We use the duality notation $\langle \alpha, f \rangle := \alpha(f)$ for $f \in \mathcal{X}$, $\alpha \in \mathcal{X}^*$. In particular, the Hahn–Banach theorem allows us to extend a nonzero continuous functional defined on a closed linear subspace of $\mathcal{X}$ to a continuous functional on the whole Banach space $\mathcal{X}$. The *adjoint operator* A^* of a densely defined linear operator A is a linear operator from $\mathcal{D}(A^*) \subset \mathcal{X}^*$ into $\mathcal{X}^*$ defined as follows. We let $\alpha \in \mathcal{D}(A^*)$ if there exists $\beta \in \mathcal{X}^*$ such that

$$\langle \alpha, Af \rangle = \langle \beta, f \rangle, \quad f \in \mathcal{D}(A), \tag{3.1}$$

in which case we set $\beta = A^*\alpha$.

We can define the Riemann integral for a function $u: [a, b] \to \mathcal{X}$. similarly to that for real-valued functions given a partition Δ of the interval $[a, b]$, i.e. $a = t_0 < t_1 < \ldots < t_n = b$ we define the Riemann sum by

$$S(\Delta, u) = \sum_{k=1}^{n} u(s_k)|t_k - t_{k-1}|$$

where $s_k \in (t_{k-1}, t_k]$. Let $f \in \mathcal{X}$. If for each $\varepsilon > 0$ there exists $\delta > 0$ such that for every partition Δ with $\sup_k |t_k - t_{k-1}| < \delta$ we have $\|f - S(P, u)\| < \varepsilon$, then the function $u \colon [a, b] \to \mathcal{X}$ is said to be Riemann integrable on the interval $[a, b]$ with the integral equal to f, which is denoted by

$$f = \int_a^b u(s)\, ds.$$

Most properties of the Riemann integral for real-valued functions can be proved for normed space-valued functions. It is easy to see that we have the following: if $u \colon [a, b] \to \mathcal{X}$ is continuous then u is uniformly continuous on $[a, b]$ and Riemann integrable on every interval $[a, t]$ for $t \in [a, b]$; in particular

$$w(t) = \int_a^t u(s)\, ds$$

is differentiable on $[a, b]$ and $w'(t) = u(t)$.

3.1.2 Definitions and Basic Properties

Let, for each $t \geq 0$, $S(t) \colon \mathcal{X} \to \mathcal{X}$ be a bounded linear operator. The family $\{S(t)\}_{t \geq 0}$ is called a *semigroup* whenever it satisfies the following conditions:

(a) $S(0) = I$, where I is the *identity operator*, i.e. $If = f$ for $f \in \mathcal{X}$,
(b) $S(t + s) = S(t)S(s)$, $s, t \geq 0$,

and it is said to be *strongly continuous* or a C_0-*semigroup* if

(c) for each $f \in \mathcal{X}$, $\|S(t)f - f\| \to 0$ as $t \downarrow 0$.

If, additionally, $\|S(t)\| \leq 1$ for every $t \geq 0$, so that each operator $S(t)$ is a contraction, then $\{S(t)\}_{t \geq 0}$ is called a *contraction semigroup*.

The *infinitesimal generator* (briefly the *generator*) of $\{S(t)\}_{t \geq 0}$ is by definition the operator A with domain $\mathscr{D}(A) \subset \mathcal{X}$ defined as

$$\mathscr{D}(A) = \{f \in \mathcal{X} \colon \lim_{t \downarrow 0} \frac{1}{t}(S(t)f - f) \text{ exists in } \mathcal{X}\},$$

$$Af = \lim_{t \downarrow 0} \frac{1}{t}(S(t)f - f), \quad f \in \mathscr{D}(A).$$

Note that $\mathscr{D}(A)$ is a linear subspace of $\mathcal{X}$ and $A \colon \mathscr{D}(A) \to \mathcal{X}$ is a linear operator.

We provide basic properties of strongly continuous semigroups and their generators. We refer to [34] for their proofs. If $\{S(t)\}_{t \geq 0}$ is a strongly continuous semigroup with generator $(A, \mathscr{D}(A))$ then the following holds:

(1) There exist constants $\omega \in \mathbb{R}$ and $M \geq 1$ such that

$$\|S(t)\| \leq M e^{\omega t}, \quad t \geq 0. \tag{3.2}$$

(2) For each $f \in \mathscr{X}$ the mapping $[0, \infty) \ni t \mapsto S(t)f \in \mathscr{X}$ is continuous.

(3) If $f \in \mathscr{X}$ then

$$\int_0^t S(s)f\,ds \in \mathscr{D}(A) \quad \text{and} \quad S(t)f - f = A\int_0^t S(s)f\,ds, \quad t > 0. \quad (3.3)$$

(4) If $f \in \mathscr{D}(A)$ then $S(t)f \in \mathscr{D}(A)$,

$$\frac{d}{dt}S(t)f = S(t)Af = AS(t)f \quad \text{and} \quad S(t)f - f = \int_0^t S(s)Af\,ds, \quad t \ge 0. \quad (3.4)$$

(5) The operator $(A, \mathscr{D}(A))$ is closed and densely defined.

We now show that the generator uniquely determines a strongly continuous semigroup. Let A_1, A_2 be two linear operators with domains $\mathscr{D}(A_1)$, $\mathscr{D}(A_2)$, respectively. We say that A_2 is an *extension* of A_1 or write $A_1 \subseteq A_2$, if $\mathscr{D}(A_1) \subseteq \mathscr{D}(A_2)$ and $A_2 f = A_1 f$ for $f \in \mathscr{D}(A_1)$.

Proposition 3.1 *Let $\{S_1(t)\}_{t \ge 0}$ and $\{S_2(t)\}_{t \ge 0}$ be strongly continuous semigroups on a Banach space $\mathscr{X}$ with generators $(A_1, \mathscr{D}(A_1))$ and $(A_2, \mathscr{D}(A_2))$, respectively. If $A_1 \subseteq A_2$ then $S_1(t)f = S_2(t)f$ for all f and $t \ge 0$.*

Proof Let $f \in \mathscr{D}(A_1)$ and $t > 0$. Define $u(s) = S_2(s)S_1(t-s)f$ for $s \in [0, t]$. The function $s \mapsto u(s)$ is differentiable at every $s \in (0, t)$ with the derivative

$$u'(s) = \frac{d}{ds}(S_2(s)S_1(t-s)f) = S_2(s)A_2 S_1(t-s)f + S_2(s)(-A_1 S_1(t-s)f).$$

We have $S_1(t-s)f \in \mathscr{D}(A_1)$ for all $t > s > 0$ and $A_2 S_1(t-s)f = A_1 S_1(t-s)f$, which shows that $u'(s) = 0$. Thus the function u is constant on $[0, t]$, which implies that $S_1(t)f = u(t) = u(0) = S_2(t)f$. Since $\mathscr{D}(A_1)$ is dense in $\mathscr{X}$ and the operators $S_1(t)$ and $S_2(t)$ are bounded, the result follows.

Remark 3.1 Suppose that $(A, \mathscr{D}(A))$ is the generator of a strongly continuous semigroup. If we let $u(t) = S(t)f$, $t \ge 0$, for a fixed $f \in \mathscr{D}(A)$ then it follows from (3.4) that $u(t) \in \mathscr{D}(A)$, the function $t \mapsto u(t)$ is differentiable with a continuous derivative, and $u(t)$ is the solution of the initial value problem

$$u'(t) = Au(t), \quad t > 0, \quad u(0) = f. \quad (3.5)$$

Similar arguments as in the proof of Proposition 3.1 show that u is the only solution of (3.5) with values in $\mathscr{D}(A)$. Consequently, for each $f \in \mathscr{D}(A)$ equation (3.5) has one and only one classical solution and it is given by $u(t) = S(t)f$ for $t \ge 0$. For $f \in \mathscr{X}$ the function $t \mapsto S(t)f$ is only a *mild solution* of (3.5), i.e. it is continuous and satisfies (3.3).

Remark 3.2 Observe that if a strongly continuous semigroup $\{S(t)\}_{t\geq 0}$ satisfies (3.2) then passing to the rescaled semigroup $T(t) = e^{-\lambda t} S(t)$, $t \geq 0$, where $\lambda \geq \omega$, yields a *bounded semigroup* on $(\mathscr{X}, \|\cdot\|)$, i.e. $\|T(t)\| \leq M$ for all $t \geq 0$. Moreover, $(A, \mathscr{D}(A))$ is the generator of $\{S(t)\}_{t\geq 0}$ if and only if $(A - \lambda I, \mathscr{D}(A))$ is the generator of $\{T(t)\}_{t\geq 0}$, since

$$\lim_{t\downarrow 0}\left[\frac{1}{t}(T(t)f - f) - \frac{1}{t}(S(t)f - f)\right] = \lim_{t\downarrow 0}\frac{1}{t}(e^{-\lambda t} - 1)S(t)f = -\lambda f.$$

Next if $\{S(t)\}_{t\geq 0}$ is a bounded semigroup on $(\mathscr{X}, \|\cdot\|)$, then we can introduce

$$\|f\|_1 = \sup\{\|S(s)f\| : s \geq 0\}, \quad f \in \mathscr{X},$$

which is a norm on $\mathscr{X}$ satisfying $\|f\| \leq \|f\|_1 \leq M\|f\|$ and $\|S(t)f\|_1 \leq \|f\|_1$ for all $t \geq 0$, $f \in \mathscr{X}$. Thus $\{S(t)\}_{t\geq 0}$ is a contraction semigroup on $(\mathscr{X}, \|\cdot\|_1)$.

3.1.3 The Resolvent

Let $(A, \mathscr{D}(A))$ be a linear operator. We say that $\lambda \in \mathbb{R}$ belongs to the *resolvent set* $\rho(A)$ of A, if the operator $\lambda I - A \colon \mathscr{D}(A) \to \mathscr{X}$ is invertible. The operator $R(\lambda, A) := (\lambda I - A)^{-1}$ for $\lambda \in \rho(A)$ is called the *resolvent operator* of A at λ.

Proposition 3.2 *Suppose that $\mu \in \rho(A)$ and that $|\lambda - \mu| < \|R(\mu, A)\|^{-1}$. Then $\lambda \in \rho(A)$ and*

$$R(\lambda, A) = \sum_{n=0}^{\infty}(\lambda - \mu)^n (R(\mu, A))^{n+1},$$

where the series converges in the operator norm.

Proof Observe that

$$R(\mu, A)R(\lambda, A) = R(\lambda, A)R(\mu, A), \quad \mu, \lambda \in \rho(A),$$

since $(\lambda I - A)(\mu I - A) = (\mu I - A)(\lambda I - A)$. We also have

$$\mu I - A = \lambda I - A + (\mu - \lambda)I = [I + (\mu - \lambda)R(\lambda, A)](\lambda I - A)$$

which implies that $R(\lambda, A)$, $\lambda \in \rho(A)$, satisfies the *resolvent identity*

$$R(\lambda, A) - R(\mu, A) = (\mu - \lambda)R(\mu, A)R(\lambda, A), \quad \mu, \lambda \in \rho(A).$$

If $\|(\lambda - \mu)R(\mu, A)\| < 1$ then the bounded operator $I - (\mu - \lambda)R(\mu, A)$ is invertible and its inverse is given by Neumann series expansion

$$(I - (\lambda - \mu)R(\mu, A))^{-1} = \sum_{n=0}^{\infty}[(\lambda - \mu)R(\mu, A)]^n,$$

this together with the resolvent identity shows that

$$R(\lambda, A) = (I - (\lambda - \mu)R(\mu, A))^{-1} R(\mu, A).$$

We now provide the integral representation of the resolvent operator of the generator of a strongly continuous semigroup. Let $\{S(t)\}_{t \geq 0}$ be a strongly continuous semigroup with generator $(A, \mathscr{D}(A))$ and let $\omega \in \mathbb{R}$ and $M \geq 1$ be constants such that (3.2) holds. For each $\lambda > \omega$ we consider

$$R(\lambda)f = \int_0^{\infty} e^{-\lambda t} S(t)f \, dt := \lim_{r \to \infty} \int_0^r e^{-\lambda t} S(t)f \, dt, \quad f \in \mathscr{X}. \qquad (3.6)$$

We will show that $R(\lambda)$ is the resolvent operator of A at λ. Since the mapping $t \mapsto e^{-\lambda t} S(t)f$ is continuous and $\|e^{-\lambda t} S(t)f\| \leq Me^{-(\lambda-\omega)t}\|f\|$ for $f \in \mathscr{X}$, we see that $R(\lambda)$ is well defined and that

$$\|R(\lambda)f\| \leq \int_0^{\infty} \|e^{-\lambda t} S(t)f\| \, dt \leq \int_0^{\infty} Me^{-(\lambda-\omega)t} \, dt\|f\| = \frac{M}{\lambda - \omega}\|f\|.$$

Thus $R(\lambda)$ is a bounded linear operator and it is the Laplace transform of the semigroup. Observe that for $h > 0$ and $f \in \mathscr{X}$ we have

$$\frac{1}{h}(S(h) - I)R(\lambda)f = \frac{1}{h}\int_0^{\infty} e^{-\lambda t}(S(h+t)f - S(t)f) \, dt$$

$$= \frac{1}{h}(e^{\lambda h} - 1)\int_0^{\infty} e^{-\lambda t} S(t)f \, dt - \frac{1}{h}e^{\lambda h}\int_0^h e^{-\lambda t} S(t)f \, dt,$$

which implies that

$$R(\lambda)f \in \mathscr{D}(A) \quad \text{and} \quad AR(\lambda)f = \lambda R(\lambda)f - f,$$

so that $(\lambda I - A)R(\lambda) = I$. Finally, if $f \in \mathscr{D}(A)$ then

$$R(\lambda)Af = \int_0^{\infty} e^{-\lambda t} S(t)Af \, dt = A\int_0^{\infty} e^{-\lambda t} S(t)f \, dt = AR(\lambda)f,$$

since (3.4) holds and the operator A is closed. Consequently, $(\omega, \infty) \subset \rho(A)$ and $R(\lambda) = R(\lambda, A)$ for $\lambda > \omega$.

We close this section with a simple result which will be useful to identify generators through their resolvents.

Lemma 3.1 *Let $(A_2, \mathscr{D}(A_2))$ be an extension of the operator $(A_1, \mathscr{D}(A_1))$. Assume that $1 \in \rho(A_1)$ and that the operator $I - A_2$ is one to one. Then $\mathscr{D}(A_2) = \mathscr{D}(A_1)$ and $A_2 = A_1$.*

Proof Let $f \in \mathscr{D}(A_2)$ and consider $g = f - A_2 f$. Since $1 \in \rho(A_1)$, the operator $(I - A_1, \mathscr{D}(A_1))$ is invertible. Thus we can find $h \in \mathscr{D}(A_1)$ such that $g = h - A_1 h$. We have $A_1 \subseteq A_2$ which implies that $f - A_2 f = h - A_1 h = h - A_2 h$. Since the operator $I - A_2$ is one to one, we conclude that $f = h$, showing that $\mathscr{D}(A_2) \subseteq \mathscr{D}(A_1)$.

3.2 Basic Examples of Semigroups

3.2.1 Uniformly Continuous Semigroups

Let A be a bounded operator on a Banach space $\mathscr{X}$. Then, for each $f \in \mathscr{X}$ and $t \geq 0$, the sequence

$$\sum_{n=0}^{m} \frac{t^n}{n!} A^n f$$

is Cauchy, thus it converges and its limit is denoted by $e^{tA} f$. The family of operators

$$S(t) = e^{tA} = \sum_{n=0}^{\infty} \frac{t^n}{n!} A^n$$

is a semigroup and it is a strongly continuous semigroup, since

$$\|S(t)f - f\| = \|\sum_{n=1}^{\infty} \frac{t^n}{n!} A^n f\| \leq \sum_{n=1}^{\infty} \frac{t^n}{n!} \|A\|^n \|f\| = (e^{t\|A\|} - 1)\|f\|.$$

Moreover, we have

$$\|S(t) - I\| \leq e^{t\|A\|} - 1,$$

which shows that the semigroup is also *uniformly continuous*, i.e.

$$\lim_{t \downarrow 0} \|S(t) - I\| = 0.$$

Observe that

$$\|S(t)f - f - tAf\| \leq \sum_{n=2}^{\infty} \frac{t^n}{n!} \|A\|^n \|f\| \leq (e^{t\|A\|} - 1 - t\|A\|)\|f\|,$$

which implies that A is the generator of the semigroup $\{S(t)\}_{t \geq 0}$.

We now show that uniformly continuous semigroups have bounded generators. Note that if a semigroup $\{S(t)\}_{t\geq 0}$ is uniformly continuous then the function $[0, \infty) \ni t \mapsto S(t) \in \mathscr{L}(\mathscr{X})$ is continuous, where the vector space $\mathscr{L}(\mathscr{X})$ of all bounded linear operators on the Banach space $\mathscr{X}$ is itself a Banach space when equipped with the operator norm. Since $S \colon [0, \infty) \to \mathscr{L}(\mathscr{X})$ is continuous, we obtain

$$\lim_{t \downarrow 0} T(t) = I, \quad \text{where } T(t) = \frac{1}{t} \int_0^t S(s)ds,$$

which implies that there is $\delta > 0$ such that $\|T(t) - I\| < 1/2$ for $0 \leq t < \delta$. Hence, the operator $T(t)$ is invertible for a sufficiently small $t > 0$. Since for any $f \in \mathscr{X}$ we have

$$\frac{1}{t}(S(t)f - f) = AT(t)f,$$

we see that the linear operator $AT(t)$ is bounded, thus

$$\|Af\| = \|AT(t)T(t)^{-1}f\| \leq \|AT(t)\| \|T(t)^{-1}\| \|f\|,$$

which shows that $\mathscr{D}(A) = \mathscr{X}$ and that A is bounded.

Remark 3.3 Note that the use of L^∞ spaces is quite limited in the theory of operator semigroups. It follows from the result of [66] that if the Banach space $\mathscr{X}$ is $L^\infty = L^\infty(X, \Sigma, m)$ and $\{S(t)\}_{t\geq 0}$ is a strongly continuous semigroup on L^∞, then the generator of $\{S(t)\}_{t\geq 0}$ is a bounded operator. Thus strongly continuous semigroups on L^∞ are necessarily uniformly continuous.

3.2.2 Multiplication Semigroups

Let (X, Σ, m) be a σ-finite measure space. Consider a measurable non-negative function $\varphi \colon X \to [0, \infty)$. Define the operator $S(t)$ on $L^1 = L^1(X, \Sigma, m)$ by

$$S(t)f(x) = e^{-t\varphi(x)}f(x).$$

We have

$$\|S(t)f\| = \int_X |e^{-t\varphi(x)}f(x)| \, m(dx) \leq \int_X |f(x)| \, m(dx) = \|f\|$$

for every $f \in L^1$ and $t \geq 0$. Thus $\{S(t)\}_{t\geq 0}$ is a semigroup on L^1. From the dominated convergence theorem it follows that

$$\lim_{t \downarrow 0} \|S(t)f - f\| = \int_X \lim_{t \downarrow 0} |e^{-t\varphi(x)}f(x) - f(x)| \, m(dx) = 0$$

for every $f \in L^1$, which implies that $\{S(t)\}_{t \geq 0}$ is a strongly continuous semigroup and it is a contraction semigroup.

We now show that the generator of $\{S(t)\}_{t \geq 0}$ is given by

$$Af = -\varphi f, \quad f \in L_\varphi^1 = \{f \in L^1 : \varphi f \in L^1\}.$$

To this end denote by $(A_1, \mathscr{D}(A_1))$ the generator of $\{S(t)\}_{t \geq 0}$. If $f \in \mathscr{D}(A_1)$ then

$$\lim_{t \downarrow 0} \frac{1}{t}(S(t)f - f) = A_1 f \in L^1,$$

thus there exists a sequence $t_n \downarrow 0$ such that

$$\lim_{n \to \infty} \frac{1}{t_n}(e^{-t_n \varphi(x)} - 1)f(x) = A_1 f(x)$$

for m-a.e. $x \in X$, which implies that

$$A_1 f(x) = -\varphi(x)f(x) \quad \text{for } m\text{-a.e. } x \in X$$

and that $\varphi f \in L^1$. Thus $\mathscr{D}(A_1) \subseteq L_\varphi^1$. Conversely, suppose that $f \in L_\varphi^1$. Since $\varphi(x) \geq 0$, we have $|e^{-t\varphi(x)} - 1| \leq t\varphi(x)$, which implies that

$$\left|\frac{1}{t}(S(t)f(x) - f(x)) + \varphi(x)f(x)\right| \leq 2\varphi(x)|f(x)|.$$

Hence, $f \in \mathscr{D}(A_1)$ and $A_1 f = -\varphi f$, since the dominated convergence theorem gives

$$\lim_{t \downarrow 0} \int \left|\frac{1}{t}(S(t)f(x) - f(x)) + \varphi(x)f(x)\right| m(dx) = 0.$$

3.2.3 Translation Semigroups

Here we give an example showing that not every semigroup is strongly continuous. Let $\mathscr{X}$ be either $L^1(\mathbb{R})$, the space of Lebesgue integrable functions on $\mathbb{R}$, or $B(\mathbb{R})$, the space of bounded functions on $\mathbb{R}$. Define the operator $S(t)$ on $\mathscr{X}$ by

$$S(t)f(x) = f(x - t), \quad x \in \mathbb{R}, \ t \geq 0.$$

It is easy to see that $\{S(t)\}_{t \geq 0}$ is a semigroup on $\mathscr{X}$.

For any $f \in \mathscr{X} = L^1(\mathbb{R})$ we have

$$\|S(t)f\| = \int_{\mathbb{R}} |f(x - t)|\, dx = \int_{\mathbb{R}} |f(x)|\, dx = \|f\|.$$

Thus $\{S(t)\}_{t\geq 0}$ is a semigroup of contractions on $L^1(\mathbb{R})$. If $f \in C_c(\mathbb{R})$, the space of continuous functions with compact support, then we have

$$\lim_{t \downarrow 0} f(x - t) = f(x)$$

for every $x \in \mathbb{R}$. By the Lebesgue dominated convergence theorem we obtain

$$\lim_{t \downarrow 0} \|S(t)f - f\| = \int_{\mathbb{R}} \lim_{t \downarrow 0} |f(x - t) - f(x)| \, dx = 0$$

for every $f \in C_c(\mathbb{R})$. Since the set $C_c(\mathbb{R})$ is a dense subset of $L^1(\mathbb{R})$, the semigroup is strongly continuous on $L^1(\mathbb{R})$.

The generator of the translation semigroup on $L^1(\mathbb{R})$ is given by $Af = -f'$ with domain

$$\mathscr{D}(A) = \{f \in L^1(\mathbb{R}) : f \text{ is absolutely continuous and } f' \in L^1(\mathbb{R})\}.$$

To see this let us denote by $(A_1, \mathscr{D}(A_1))$ the generator of the translation semigroup $\{S(t)\}_{t\geq 0}$. Recall that we have $S(t)f(x) = f(x - t)$, $x \in \mathbb{R}$, $t \geq 0$, $f \in L^1(\mathbb{R})$. First take $f \in \mathscr{D}(A_1)$ so that

$$\lim_{t \downarrow 0} \frac{1}{t}(S(t)f - f) = g \in L^1(\mathbb{R}).$$

Then for every compact interval $[a, b] \subset \mathbb{R}$ we have

$$\left| \int_{[a,b]} \frac{1}{t}(f(x - t) - f(x)) \, dx - \int_{[a,b]} g(x) \, dx \right| \leq \left\| \frac{1}{t}(S(t)f - f) - g \right\|,$$

which implies that

$$\lim_{t \downarrow 0} \int_{[a,b]} \frac{f(x - t) - f(x)}{t} \, dx = \int_{[a,b]} g(x) \, dx.$$

For all sufficiently small $t > 0$ we have

$$\int_{[a,b]} \frac{1}{t}(f(x - t) - f(x)) \, dx = \frac{1}{t} \int_{[a-t,a]} f(x) \, dx - \frac{1}{t} \int_{[b-t,b]} f(x) \, dx.$$

Since

$$\lim_{t \downarrow 0} \frac{1}{t} \int_{[b-t,b]} f(x) \, dx = f(b)$$

for a.e. $b \in \mathbb{R}$, we conclude that

$$\int_{[a,b]} g(x)\,dx = f(a) - f(b)$$

for a.e. $a, b \in \mathbb{R}$. Thus f is absolutely continuous and its derivative being equal to $-g$ in the $L^1(\mathbb{R})$ space is integrable, which shows that the operator $(A, \mathscr{D}(A))$ is an extension of the generator $(A_1, \mathscr{D}(A_1))$. We have $1 \in \rho(A_1)$. Note that the general solution of the differential equation $f(x) + f'(x) = 0$ is of the form $f(x) = ce^{-x}$ for $x \in \mathbb{R}$, where c is a constant. Thus the operator $I - A$ is one to one which implies that $A = A_1$ by Lemma 3.1.

Consider now the semigroup $\{S(t)\}_{t \geq 0}$ on the space $\mathscr{X} = B(\mathbb{R})$. Then $\{S(t)\}_{t \geq 0}$ is a semigroup of contractions, since

$$\|S(t)f\|_u = \sup_{x \in \mathbb{R}} |S(t)f(x)| \leq \sup_{x \in \mathbb{R}} |f(x)| = \|f\|_u.$$

However, it is no longer strongly continuous on the space $B(\mathbb{R})$. To see this take $f(x) = \mathbf{1}_{[0,1)}(x)$, $x \in \mathbb{R}$. We have

$$|S(t)f(x) - f(x)| = |\mathbf{1}_{[t,1+t)}(x) - \mathbf{1}_{[0,1)}(x)| = |\mathbf{1}_{[1,1+t)}(x) - \mathbf{1}_{[0,t)}(x)|$$

for $t \in (0, 1)$ and $x \in \mathbb{R}$, which implies that

$$\|S(t)f - f\|_u = \sup_{x \in \mathbb{R}} |S(t)f(x) - f(x)| = 1 \quad \text{for all } t \in (0, 1).$$

Remark 3.4 Note that if $\{S(t)\}_{t \geq 0}$ is a semigroup on a Banach space $\mathscr{X}$, then

$$\mathscr{X}_0 = \{f \in \mathscr{X} : \lim_{t \downarrow 0} \|S(t)f - f\| = 0\}$$

is a closed linear subspace of $\mathscr{X}$ and $S(t)(\mathscr{X}_0) \subseteq \mathscr{X}_0$ for all $t \geq 0$. Thus $\{S(t)\}_{t \geq 0}$ is a strongly continuous semigroup on the Banach space $\mathscr{X}_0$.

For the translation semigroup on $\mathscr{X} = B(\mathbb{R})$ it is easy to see that the subspace $\mathscr{X}_0$ contains all uniformly continuous functions on $\mathbb{R}$.

3.3 Generators of Contraction Semigroups

3.3.1 The Hille–Yosida Theorem

The abstract theory of contraction semigroups was developed independently by Hille [49] and Yosida [120]. The first major result in the semigroup theory is the following.

Theorem 3.1 (Hille–Yosida) *A linear operator $(A, \mathscr{D}(A))$ on a Banach space $\mathscr{X}$ is the generator of a contraction semigroup if and only if $\mathscr{D}(A)$ is dense in $\mathscr{X}$, the* *resolvent set $\rho(A)$ of A contains $(0, \infty)$, and for every $\lambda > 0$*

$$\|\lambda R(\lambda, A)\| \leq 1.$$

In that case, the semigroup $\{S(t)\}_{t \geq 0}$ with generator $(A, \mathscr{D}(A))$ is given by

$$S(t)f = \lim_{\lambda \to \infty} e^{-\lambda t} \sum_{n=0}^{\infty} \frac{\lambda^n t^n}{n!} (\lambda R(\lambda, A))^n f, \quad f \in \mathscr{X}, t \geq 0. \qquad (3.7)$$

For a proof see [34] or [35]. To check that an operator A with a dense domain is the generator by using the Hille–Yosida theorem, we need to show that for each $\lambda > 0$ and $g \in \mathscr{X}$ there exists a unique solution $f \in \mathscr{D}(A)$ of

$$\lambda f - Af = g \quad \text{and} \quad \lambda\|f\| \leq \|g\|.$$

The following is a key notion towards a characterization of the generator of a contraction semigroup that does not require explicit knowledge of the resolvent operator. An operator $(A, \mathscr{D}(A))$ is called *dissipative*, if

$$\|\lambda f - Af\| \geq \lambda\|f\| \quad \text{for all } f \in \mathscr{D}(A) \text{ and } \lambda > 0. \qquad (3.8)$$

Observe that if $(A, \mathscr{D}(A))$ is the generator of a strongly continuous contraction semigroup then $\lambda R(\lambda, A) = (I - \lambda^{-1}A)^{-1}$ is a contraction for each $\lambda > 0$. Thus if for a given $f \in \mathscr{D}(A)$ and $\lambda > 0$ we let $g = f - \lambda^{-1}Af$ then $f = (I - \lambda^{-1}A)^{-1}g$ and we have $\|f - \lambda^{-1}Af\| = \|g\| \geq \|(I - \lambda^{-1}A)^{-1}g\| = \|f\|$, which shows that A is dissipative. Also the converse is true, as we prove next.

Lemma 3.2 *Suppose that $(A, \mathscr{D}(A))$ is dissipative. Then for each $\lambda > 0$ the operator $I - \lambda^{-1}A$ is one to one and its inverse $(I - \lambda^{-1}A)^{-1} \colon \mathscr{R}(\lambda I - A) \to \mathscr{D}(A)$ is a contraction, where $\mathscr{R}(\lambda I - A)$ is the range of $\lambda I - A$. Moreover, if $\mathscr{R}(\lambda I - A) = \mathscr{X}$ for some $\lambda > 0$ then $\mathscr{R}(\lambda I - A) = \mathscr{X}$ for all $\lambda > 0$ and $(0, \infty) \subseteq \rho(A)$.*

Proof The linear operator $\lambda I - A$ is one to one for each $\lambda > 0$, since from $0 = \|\lambda f - Af\| \geq \lambda\|f\|$, it follows that $\|f\| = 0$. For $g \in \mathscr{R}(I - \lambda^{-1}A)$ we let $f = (I - \lambda^{-1}A)^{-1}g$ which gives

$$\|g\| = \|(I - \lambda^{-1}A)(I - \lambda^{-1}A)^{-1}g\| \geq \|(I - \lambda^{-1}A)^{-1}g\|.$$

Hence, $(I - \lambda^{-1}A)^{-1}$ is a contraction.

Suppose now that $(\lambda_0 I - A)(\mathscr{D}(A)) = \mathscr{X}$ for some $\lambda_0 > 0$. Then the operator $\lambda_0 I - A$ defined on $\mathscr{X}$ is invertible and $\lambda_0 \in \rho(A)$. We have $\|R(\lambda_0, A)\|^{-1} \geq \lambda_0$. From Proposition 3.2 it follows that each λ satisfying

$$|\lambda - \lambda_0| < \lambda_0$$

belongs to $\rho(A)$, which implies that $(\lambda I - A)(\mathscr{D}(A)) = \mathscr{X}$ for $\lambda \in (0, 2\lambda_0)$. Repeating the argument we conclude that $(\lambda I - A)(\mathscr{D}(A)) = \mathscr{X}$ for all $\lambda > 0$.

We come to the key result, the Lumer–Phillips reformulation [67] of the Hille–Yosida theorem.

Theorem 3.2 *A linear operator $(A, \mathscr{D}(A))$ on a Banach space $\mathscr{X}$ is the generator of a contraction semigroup if and only if $\mathscr{D}(A)$ is dense in $\mathscr{X}$, $(A, \mathscr{D}(A))$ is dissipative, and the range of the operator $\lambda I - A$ is $\mathscr{X}$ for some $\lambda > 0$.*

To check that a densely defined operator A is the generator by using this reformulation of the Hille–Yosida theorem, we need to show that A is dissipative and that there is $\lambda > 0$ such that for each $g \in \mathscr{X}$ there exists a solution $f \in \mathscr{D}(A)$ of the equation

$$\lambda f - Af = g.$$

We now provide an equivalent definition of a dissipative operator. Let $\mathscr{X}^*$ be the dual space of $\mathscr{X}$. A linear operator $(A, \mathscr{D}(A))$ is dissipative if and only if for each $f \in \mathscr{D}(A)$ there exists $\alpha \in \mathscr{X}^*$ with $\|\alpha\| \leq 1$ such that

$$\langle \alpha, f \rangle = \|f\| \quad \text{and} \quad \langle \alpha, Af \rangle \leq 0. \tag{3.9}$$

It is easy to prove the "if" part, since for each $\lambda > 0$ and $f \in \mathscr{D}(A)$ condition (3.9) implies

$$\lambda \|f\| = \langle \alpha, \lambda f \rangle \leq \langle \alpha, \lambda f \rangle - \langle \alpha, Af \rangle = \langle \alpha, \lambda f - Af \rangle \leq \|\alpha\| \|\lambda f - Af\|.$$

For the proof of the converse see, e.g. [34, Proposition II.3.23].

It follows from the Hahn–Banach theorem that the normalized duality set

$$J(f) = \{\alpha \in \mathscr{X}^* : \|\alpha\| \leq 1 \text{ and } \langle \alpha, f \rangle = \|f\|\}$$

is not empty. In particular, if P is a contraction then the operator $A = P - I$ is dissipative, since for any $f \in \mathscr{X}$ and $\alpha \in J(f)$ we have

$$\langle \alpha, Af \rangle = \langle \alpha, Pf - f \rangle = \langle \alpha, Pf \rangle - \langle \alpha, f \rangle \leq \|\alpha\| \|f\| - \|f\| \leq 0.$$

Also the operator $A = \lambda(P - I)$ is dissipative for every $\lambda > 0$.

We have the following extension of the Hille–Yosida theorem to arbitrary semigroups, as discovered independently by Feller, Phillips and Miyadera (see [34]).

Theorem 3.3 *A linear operator $(A, \mathscr{D}(A))$ generates a strongly continuous semigroup on $\mathscr{X}$ satisfying (3.2) with constants $M \geq 1$ and $\omega \in \mathbb{R}$ if and only if $\mathscr{D}(A)$ is dense in $\mathscr{X}$, the resolvent set $\rho(A)$ of A contains (ω, ∞), and*

$$\|[(\lambda - \omega)R(\lambda, A)]^n\| \leq M, \quad \lambda > \omega, \quad n = 1, 2, \ldots. \tag{3.10}$$

An operator A with resolvent operator $R(\lambda, A)$ satisfying (3.10) for some constants M, ω is called a *Hille–Yosida operator*.

Remark 3.5 If an operator A is not densely defined, then we can still generate a semigroup on a closed subspace of the Banach space $\mathscr{X}$. In particular we have the following; see [34, Chapter III]. If an operator $(A, \mathscr{D}(A))$ is dissipative and the range of the operator $\lambda I - A$ is $\mathscr{X}$ for some $\lambda > 0$, then the part $A_|$ of the operator A in the subspace $\mathscr{X}_0 = \overline{\mathscr{D}(A)}$, i.e.

$$A_| f = Af \quad \text{for} \quad f \in \mathscr{D}(A_|) = \{f \in \mathscr{D}(A) \cap \mathscr{X}_0 : Af \in \mathscr{X}_0\},$$

is densely defined and it generates a contraction semigroup on $\mathscr{X}_0$. If $(A, \mathscr{D}(A))$ is a Hille–Yosida operator then it generates a strongly continuous semigroup on $\mathscr{X}_0$.

3.3.2 The Lumer–Phillips Theorem

An alternative form of the Lumer–Phillips theorem is more readily applicable. To state it we need to introduce the following concepts. The operator A is said to be *closable*, if it has an extension which is a closed operator. Another way to state this is that the closure in $\mathscr{X} \times \mathscr{X}$ of the graph

$$\mathscr{G}(A) = \{(f, g) \in \mathscr{X} \times \mathscr{X} : f \in \mathscr{D}(A), \; g = Af\}$$

of A is a graph of a linear operator, i.e. $(0, g) \in \overline{\mathscr{G}(A)}$ implies that $g = 0$. If A is closable, then the *closure* $\overline{A}$ *of* A is the closed operator whose graph is the closure of the graph of A. We have

$$\mathscr{D}(\overline{A}) = \{f \in \mathscr{X} : \text{there exist } f_n \in \mathscr{D}(A) \text{ and } g \in \mathscr{X} \text{ such that } f_n \to f \text{ and } Af_n \to g\}$$

and $\overline{A}f = g$. It is easily seen that if $(A, \mathscr{D}(A))$ is dissipative and with dense domain, then it is closable and its closure $(\overline{A}, \mathscr{D}(\overline{A}))$ is dissipative and the range of $\lambda I - \overline{A}$ is the closure of the range of $\lambda I - A$. This gives an extension of the Hille–Yosida theorem.

Theorem 3.4 (Lumer–Phillips) *A linear operator* $(A, \mathscr{D}(A))$ *on a Banach space* $\mathscr{X}$ *is closable and its closure is the generator of a contraction semigroup if and only if* $\mathscr{D}(A)$ *is dense in* $\mathscr{X}$, $(A, \mathscr{D}(A))$ *is dissipative, and the range of the operator* $\lambda I - A$ *is dense in* $\mathscr{X}$ *for some* $\lambda > 0$.

We now show how to check the range condition from the Lumer–Phillips theorem using the notion of the adjoint operator. It is easy to see that if A is a densely defined linear operator and $\lambda \in \rho(A)$ then $\lambda \in \rho(A^*)$ and $R(\lambda, A)^* = R(\lambda, A^*)$, where A^* is the adjoint of A. Using the Lumer–Phillips theorem we now prove the following

Corollary 3.1 *Let $(A, \mathscr{D})$ be a dissipative operator with dense domain $\mathscr{D}$. Then the closure of $(A, \mathscr{D})$ is the generator of a contraction semigroup if and only if the operator $\lambda I^* - A^*$ is one to one for some $\lambda > 0$, where I^* is the identity in $\mathscr{X}^*$.*

Proof Suppose that the closure of the set $(\lambda I - A)\mathscr{D}$ is not equal to $\mathscr{X}$. From the Hahn–Banach theorem it follows that there must exist $\alpha \in \mathscr{X}^*, \alpha \neq 0$, such that

$$\langle \alpha, \lambda f - Af \rangle = 0$$

for all $f \in \mathscr{D}$. This implies that $\alpha \in \mathscr{D}(A^*)$ and $A^*\alpha = \lambda\alpha$. Thus $\lambda I^* - A^*$ is not one to one.

In general, it is difficult to identify the whole domain of the generator. The following concept is useful. If $(A, \mathscr{D}(A))$ is a closed linear operator then a linear subspace $\mathscr{D}$ of $\mathscr{D}(A)$ is called a *core* for A if the closure of the restriction of A to $\mathscr{D}$, denoted by $A|_{\mathscr{D}}$, is equal to A; in symbols, $\overline{A|_{\mathscr{D}}} = A$. We give a sufficient condition for $\mathscr{D}$ to be a core.

Theorem 3.5 *Let $\{S(t)\}_{t\geq 0}$ be a contraction semigroup with generator $(A, \mathscr{D}(A))$. Suppose that $\mathscr{D}_0 \subseteq \mathscr{D} \subseteq \mathscr{D}(A)$ are subspaces of $\mathscr{X}$ such that $S(t)(\mathscr{D}_0) \subseteq \mathscr{D}$ for all $t > 0$. Then $\mathscr{D}_0 \subseteq \mathscr{R}(\lambda I - A|_{\mathscr{D}})$ for all $\lambda > 0$. In particular, if $\mathscr{D}_0$ is dense in $\mathscr{X}$ then $\mathscr{D}$ is a core for $(A, \mathscr{D}(A))$.*

Proof Let $g \in \mathscr{D}_0$ and $\mu > 0$. The operator $(\mu A, \mathscr{D}(A))$ is the generator of the rescaled semigroup $\{S(\mu t)\}_{t\geq 0}$. From (3.6) it follows that we have

$$g = (I - \mu A)^{-1}(I - \mu A)g = \lim_{r\to\infty} \int_0^r e^{-t} S(\mu t)(I - \mu A)g \, dt$$

$$= \lim_{r\to\infty} \lim_{n\to\infty} \frac{r}{n} \sum_{j=0}^{n-1} e^{-jr/n} S(\mu j r/n)(I - \mu A)g,$$

which implies the first claim, since $S(\mu t)(I - \mu A)g = (I - \mu A)S(\mu t)g \in \mathscr{R}(I - \mu A|_{\mathscr{D}})$ for any $t \geq 0$. The second claim follows from the Lumer–Phillips theorem and Lemma 3.1.

Finally, the following result gives the uniqueness of the semigroup generated by an extension of a given operator.

Theorem 3.6 *Let $\{S(t)\}_{t\geq 0}$ be a strongly continuous semigroup with generator extending an operator $(A, \mathscr{D})$. The following conditions are equivalent:*

(1) $\mathscr{D}$ is a core for the generator of $\{S(t)\}_{t\geq 0}$.
(2) The closure of $(A, \mathscr{D})$ is the generator of a strongly continuous semigroup.
(3) The semigroup $\{S(t)\}_{t\geq 0}$ is the only semigroup with generator being an extension of the operator $(A, \mathscr{D})$.

Proof Let A_1 be a closed extension of $(A, \mathscr{D})$. Since the closure $\overline{A}$ of $(A, \mathscr{D})$ is the smallest closed extension of $(A, \mathscr{D})$, we have $\overline{A} \subseteq A_1$. If $\overline{A}$ and A_1 are generators of strongly continuous semigroups, then the semigroups are equal. If $\mathscr{D}$ is not a core for the generator of $\{S(t)\}_{t \geq 0}$, then one can find an infinite number of extensions of A which are generators (see [5, A-II, Theorem 1.33]).

3.3.3   Perturbations of Semigroups

An important tool for the construction of a semigroup is perturbation theory; see [34, Chapter III]. Here we only give the Phillips perturbation theorem which concerns bounded perturbations and the variation of parameters formula. First, we need the following result.

Lemma 3.3 *Assume that $(A, \mathscr{D}(A))$ and $(B, \mathscr{D}(B))$ are linear operators such that $\mathscr{D}(A) \subseteq \mathscr{D}(B)$. Let $\lambda \in \rho(A)$ and let $BR(\lambda, A)$ be a bounded operator on $\mathscr{X}$. Then the operator $(A + B, \mathscr{D}(A))$ satisfies*

$$\mathscr{R}(\lambda I - A - B) = \mathscr{R}(I - BR(\lambda, A)).$$

Moreover, $\lambda \in \rho(A + B)$ if and only if $1 \in \rho(BR(\lambda, A))$. In that case,

$$R(\lambda, A + B) = R(\lambda, A)(I - BR(\lambda, A))^{-1}.$$

Proof Since $\lambda \in \rho(A)$, we have $\mathscr{R}(R(\lambda, A)) = \mathscr{D}(A)$ and $(\lambda I - A)R(\lambda, A)f = f$ for $f \in \mathscr{X}$, which gives

$$(\lambda I - A - B)R(\lambda, A)f = (I - BR(\lambda, A))f \quad \text{for all } f \in \mathscr{X}. \tag{3.11}$$

Thus, the first assertion follows. We also have

$$(\lambda I - A - B) = (I - BR(\lambda, A))(\lambda I - A),$$

this together with (3.11) proves the claim.

We now use the Hille–Yosida theorem to prove the following theorem.

Theorem 3.7 (Phillips Perturbation) *Let $(A, \mathscr{D}(A))$ be the generator of a strongly continuous semigroup and let B be a bounded operator. Then $(A + B, \mathscr{D}(A))$ is the generator of a strongly continuous semigroup.*

Proof We can assume that $(A, \mathscr{D}(A))$ is the generator of a contraction semigroup. Then for each $\lambda > 0$ the operator $BR(\lambda, A)$ is bounded and $\|BR(\lambda, A)\| \leq \|B\|/\lambda$. If we take $\lambda > \|B\|$, so that $1 \in \rho(BR(\lambda, A))$, then it follows from Lemma 3.3 that $\lambda \in \rho(A + B)$ and

$$\|R(\lambda, A + B)\| \le \frac{1}{\lambda} \frac{1}{1 - \|B\|/\lambda} = \frac{1}{\lambda - \|B\|}.$$

Since $R(\lambda, A + B) = R(\lambda - \|B\|, A + B - \|B\|I)$, we conclude that $(0, \infty) \subset \rho(A + B - \|B\|I)$ and $\|\mu R(\mu, A + B - \|B\|I)\| \le 1$. From the Hille–Yosida theorem it follows that the operator $(A + B - \|B\|I, \mathcal{D}(A))$ is the generator of a contraction semigroup. By rescaling, the operator $(A + B, \mathcal{D}(A))$ is also the generator.

We now consider two strongly continuous semigroups $\{S(t)\}_{t \ge 0}$ and $\{P(t)\}_{t \ge 0}$ with generators $(A, \mathcal{D}(A))$ and $(C, \mathcal{D}(C))$. Suppose that C is an extension of the operator $(A + B, \mathcal{D}(A))$ for some operator $(B, \mathcal{D}(B))$ with $\mathcal{D}(B) \supseteq \mathcal{D}(A)$. Then the semigroup $\{P(t)\}_{t \ge 0}$ satisfies the integral equation

$$P(t)f = S(t)f + \int_0^t P(t - s)BS(s)f\, ds \tag{3.12}$$

for any $f \in \mathcal{D}(A)$ and $t \ge 0$. To see this observe that for $f \in \mathcal{D}(A)$ the function $u(s) = P(t - s)S(s)f$ is continuously differentiable with derivative

$$u'(s) = P(t - s)AS(s)f - P(t - s)CS(s)f = -P(t - s)BS(s)f,$$

since $S(s)f \in \mathcal{D}(A)$ and $CS(s)f = AS(s)f + BS(s)f$. This implies that

$$P(t)f - S(t)f = u(0) - u(t) = -\int_0^t u'(s)ds = \int_0^t P(t - s)BS(s)f\, ds.$$

If the operator B is bounded then the semigroup $\{P(t)\}_{t \ge 0}$ is also given by the *Dyson–Phillips expansion*

$$P(t)f = \sum_{n=0}^{\infty} S_n(t)f, \tag{3.13}$$

where $S_0(t)f = S(t)f$ and

$$S_{n+1}(t)f = \int_0^t S_n(t - s)BS_0(s)f\, ds, \quad f \in \mathcal{D}(A), \ n \ge 0. \tag{3.14}$$

If B is bounded then equalities (3.12) and (3.14) can be extended to all $f \in \mathcal{X}$. Moreover, instead of (3.12) one can also consider the variation of parameters formula

$$P(t)f = S(t)f + \int_0^t S(t - s)BP(s)f\, ds.$$

We close this section by giving the Phillips perturbation theorem for perturbations of contraction semigroups.

Corollary 3.2 *Let $\bar{P}$ be a contraction on $\mathscr{X}$ and let $(A_0, \mathscr{D}(A_0))$ be the generator of a contraction semigroup $\{P_0(t)\}_{t\geq 0}$. If $\lambda > 0$ is a constant then the operator $(A_0 - \lambda I + \lambda \bar{P}, \mathscr{D}(A_0))$ is the generator of a contraction semigroup $\{P(t)\}_{t\geq 0}$ and*

$$P(t) = e^{-\lambda t} \sum_{n=0}^{\infty} \lambda^n S_n(t) f, \quad f \in L^1, \ t \geq 0, \tag{3.15}$$

where S_n are as in (3.14) with $S_0(t) = P_0(t)$ and $B = \bar{P}$.

3.3.4 Perturbing Boundary Conditions

In this section, we describe, based on [44, 45], a perturbation method related to operators with boundary conditions. On a Banach space $\mathscr{X}$ we consider a linear operator $(A, \mathscr{D})$, called the maximal operator in the sense that it has a sufficiently large domain $\mathscr{D} \subset \mathscr{X}$. We assume that there is a second Banach space $\partial \mathscr{X}$ which will serve here as the boundary space. We consider two operators $\Psi_0, \Psi: \mathscr{D} \to \partial \mathscr{X}$, called boundary operators, and we define

$$\mathscr{D}(A) = \{f \in \mathscr{D}: \Psi_0(f) = \Psi(f)\}. \tag{3.16}$$

We denote by $(A_0, \mathscr{D}(A_0))$ the restriction of the operator $(A, \mathscr{D})$ to the zero boundary condition

$$A_0 f = Af, \quad f \in \mathscr{D}(A_0) = \{f \in \mathscr{D}: \Psi_0(f) = 0\}. \tag{3.17}$$

We now describe how we can rewrite the operator $(A, \mathscr{D}(A))$ using perturbations.
We define two operators $\mathscr{A}, \mathscr{B}: \mathscr{D}(\mathscr{A}) \to \mathscr{X} \times \partial \mathscr{X}$ with $\mathscr{D}(\mathscr{A}) = \mathscr{D} \times \{0\}$ by

$$\mathscr{A}(f, 0) = (Af, -\Psi_0 f) \quad \text{and} \quad \mathscr{B}(f, 0) = (0, \Psi f) \quad \text{for } f \in \mathscr{D}.$$

We have

$$(\mathscr{A} + \mathscr{B})(f, 0) = (Af, \Psi f - \Psi_0 f), \quad f \in \mathscr{D}.$$

The part of $(\mathscr{A}, \mathscr{D}(\mathscr{A}))$ in $\mathscr{X} \times \{0\}$ denoted by $(\mathscr{A}_|, \mathscr{D}(\mathscr{A}_|))$, i.e.

$$\mathscr{D}(\mathscr{A}_|) = \{(f, 0) \in \mathscr{D}(\mathscr{A}) \cap \mathscr{X} \times \{0\}: \mathscr{A}(f, 0) \in \mathscr{X} \times \{0\}\},$$

can be identified with $(A_0, \mathscr{D}(A_0))$; we have

$$\mathscr{D}(\mathscr{A}_|) = \mathscr{D}(A_0) \times \{0\}, \quad \mathscr{A}_|(f, 0) = (A_0 f, 0).$$

Hence, $(\mathscr{A}_|, \mathscr{D}(\mathscr{A}_|))$ is the generator of a strongly continuous semigroup on $\mathscr{X} \times \{0\}$ if and only if $(A_0, \mathscr{D}(A_0))$ is the generator of a strongly continuous semigroup on $\mathscr{X}$. Similarly, the part of $(\mathscr{A}+\mathscr{B}, \mathscr{D}(\mathscr{A}))$ in $\mathscr{X} \times \{0\}$, denoted by $(\mathscr{A}+\mathscr{B})_|$, can be identified with $(A, \mathscr{D}(A))$; we have

$$\mathscr{D}((\mathscr{A} + \mathscr{B})_|) = \mathscr{D}(A) \times \{0\}, \quad (\mathscr{A} + \mathscr{B})_|(f, 0) = (Af, 0), \quad f \in \mathscr{D}(A).$$

Using ideas of Greiner [44] we are able to compute the resolvent operator of the operator $(A, \mathscr{D}(A))$ as stated next.

Lemma 3.4 *Let $(A_0, \mathscr{D}(A_0))$ be as in (3.17) and let $\lambda \in \rho(A_0)$. Assume that the operator $\Psi_0 \colon \mathscr{D} \to \partial \mathscr{X}$ restricted to the nullspace $\mathscr{N}(\lambda I - A) = \{f \in \mathscr{D} \colon \lambda f - Af = 0\}$ is invertible with bounded inverse $\Psi(\lambda) \colon \partial \mathscr{X} \to \mathscr{N}(\lambda I - A)$. For the operator $(A, \mathscr{D}(A))$ with $\mathscr{D}(A)$ as in (3.16) we have $\lambda \in \rho(A)$ if and only if $I_{\partial \mathscr{X}} - \Psi\Psi(\lambda)$ is invertible, where $I_{\partial \mathscr{X}}$ is the identity operator on $\partial \mathscr{X}$. In that case, if $\Psi R(\lambda, A_0)$ and $\Psi\Psi(\lambda)$ are bounded then the resolvent operator of A at λ is given by*

$$R(\lambda, A)f = (I + \Psi(\lambda)(I_{\partial \mathscr{X}} - \Psi\Psi(\lambda))^{-1}\Psi)R(\lambda, A_0)f, \quad f \in \mathscr{X}. \quad (3.18)$$

Proof We show that the resolvent of the operator $\mathscr{A}$ at λ is given by

$$R(\lambda, \mathscr{A})(f, f_\partial) = (R(\lambda, A_0)f + \Psi(\lambda)f_\partial, 0), \quad (f, f_\partial) \in \mathscr{X} \times \partial \mathscr{X}.$$

Let $\mathscr{R}_\lambda(f, f_\partial) = (R(\lambda, A_0)f + \Psi(\lambda)f_\partial, 0)$ for $(f, f_\partial) \in \mathscr{X} \times \partial \mathscr{X}$. Observe that $(\lambda - \mathscr{A})\mathscr{R}_\lambda(f, f_\partial)$ is equal to

$$((\lambda - A)R(\lambda, A_0)f + (\lambda - A)\Psi(\lambda)f_\partial, \Psi_0 R(\lambda, A_0)f + \Psi_0\Psi(\lambda)f_\partial).$$

Since $R(\lambda, A_0)f \in \mathscr{D}(A_0)$ and $\Psi(\lambda)f_\partial \in \mathscr{N}(\lambda I - A)$, we obtain

$$AR(\lambda, A_0)f = A_0 R(\lambda, A_0)f, \quad \Psi_0 R(\lambda, A_0)f = 0, \quad \text{and} \quad (\lambda - A)\Psi(\lambda)f_\partial = 0.$$

Thus,

$$(\lambda - \mathscr{A})\mathscr{R}_\lambda(f, f_\partial) = ((\lambda - A_0)R(\lambda, A_0)f, \Psi_0\Psi(\lambda)f_\partial) = (f, f_\partial).$$

Similarly, for $f \in \mathscr{D}(A)$ and $f_\partial = 0$ we have

$$\mathscr{R}_\lambda(\lambda - \mathscr{A})(f, f_\partial) = (R(\lambda, A_0)(\lambda I - A)f + \Psi(\lambda)\Psi_0 f, 0) = (f, f_\partial),$$

since $f = f - \Psi(\lambda)\Psi_0 f + \Psi(\lambda)\Psi_0 f$ with $f - \Psi(\lambda)\Psi_0 f \in \mathscr{D}(A)$ and $\Psi(\lambda)\Psi_0 f \in \mathscr{N}(\lambda I - A)$ implying that

$$R(\lambda, A_0)(\lambda I - A)f = R(\lambda, A_0)(\lambda - A_0)(f - \Psi(\lambda)\Psi_0 f) = f - \Psi(\lambda)\Psi_0 f.$$

Observe that $I - \mathscr{B}R(\lambda, \mathscr{A})$ is invertible if and only if $I - \Psi\Psi(\lambda)$ is invertible, since

$$(I - \mathscr{B}R(\lambda, \mathscr{A}))(f, f_\partial) = (f, -\Psi R(\lambda, A_0)f + (I - \Psi\Psi(\lambda))f_\partial)$$

for any $(f, f_\partial) \in \mathscr{X} \times \partial\mathscr{X}$. The resolvent of $\mathscr{A} + \mathscr{B}$ is, by Lemma 3.3, equal to

$$R(\lambda, \mathscr{A} + \mathscr{B}) = R(\lambda, \mathscr{A})(I - \mathscr{B}R(\lambda, \mathscr{A}))^{-1},$$

which completes the proof.

Remark 3.6 It is assumed in [44] that the operators $(A, \mathscr{D})$ and $(\Psi_0, \mathscr{D})$ are closed, the operator Ψ is bounded, the range of the operator Ψ_0 is equal to $\partial\mathscr{X}$, and that there are constants $\gamma > 0$ and $\lambda_0 \in \mathbb{R}$ such that

$$\|\Psi_0(f)\| \geq \lambda\gamma\|f\|, \quad f \in \mathscr{N}(\lambda I - A), \ \lambda > \lambda_0. \tag{3.19}$$

These conditions imply that $\|\lambda\Psi(\lambda)\| \leq \gamma^{-1}$ for all $\lambda \in \rho(A_0)$ and that one can apply the Hille–Yosida theorem to show that the operator A generates a strongly continuous semigroup. In fact, operators $I - \Psi(\lambda)\Psi$ and $I - \Psi\Psi(\lambda)$ are invertible and $(I - \Psi(\lambda)\Psi)^{-1} = I + \Psi(\lambda)(I - \Psi\Psi(\lambda))\Psi$. In Sect. 4.1.5 we shall provide an extension of Grainer's result to unbounded perturbations in L^1 space.

Chapter 4
Stochastic Semigroups

In this chapter, we introduce stochastic semigroups as strongly continuous semi-groups of stochastic operators on L^1 spaces. We provide characterizations of their generators and we explain their connection with PDMPs as defined in the previous chapters. We give examples of such semigroups which correspond to pure jump-type processes, to deterministic processes, to semiflows with jumps and to randomly switched dynamical systems.

4.1 Aspects of Positivity

4.1.1 Positive Operators

In this chapter, we assume that the Banach space $\mathscr{X}$ is $L^1 = L^1(X, \Sigma, m)$, where (X, Σ, m) is a σ-finite measure space, with the norm

$$\|f\| = \int_X |f(x)| \, m(dx), \quad f \in L^1.$$

We can write any f as the difference of two non-negative functions $f = f^+ - f^-$, where the *positive part* f^+ and the *negative part* f^- are defined by

$$f^+ = \max\{0, f\} \quad \text{and} \quad f^- = (-f)^+ = \max\{0, -f\}.$$

We define the positive cone L^1_+ to be the set of $f \in L^1$ which are positive $f \geq 0$. A linear operator $A: \mathscr{D}(A) \to L^1$ is said to be *positive* if $Af \geq 0$ for $f \in \mathscr{D}(A)_+$, where $\mathscr{D}(A)_+ = \mathscr{D}(A) \cap L^1_+$, and we write $A \geq 0$. A positive and everywhere defined oper-

© The Author(s) 2017
R. Rudnicki and M. Tyran-Kamińska, *Piecewise Deterministic Processes
in Biological Models*, SpringerBriefs in Mathematical Methods,
DOI 10.1007/978-3-319-61295-9_4

ator is a bounded operator and its norm is determined through values on the positive
cone, as we show next.

Proposition 4.1 *Let A be a linear operator with $\mathscr{D}(A) = L^1$. If*

$$\|(Af)^+\| \leq \|f^+\|, \quad f \in L^1,$$

then A is positive. If A is positive then A is a bounded operator and

$$\|A\| = \sup_{f \geq 0, \|f\|=1} \|Af\|.$$

Proof We have $(Af)^- = (-Af)^+ = (A(-f))^+$. Thus, if $f \geq 0$ then $f^- = 0$ and

$$\|(Af)^-\| = \|(A(-f))^+\| \leq \|(-f)^+\| = \|f^-\| = 0.$$

This shows that $Af = (Af)^+ \geq 0$.

Observe that we have $-|f| \leq f \leq |f|$ for any $f \in L^1$, thus $|Af| \leq A|f|$, which
implies that $\|Af\| \leq \|A|f|\|$. If we let a be equal to the supremum in the right-hand
of the equality, then $a \leq \|A\|$. Since $|f| \geq 0$, we see that $\|Af\| \leq \|A|f|\| \leq a\|f\|$
for any $f \in L^1$, which shows that $\|A\| = a$

Suppose that A is not bounded. Then we can find a sequence $f_n \in L^1_+$ such that
$\|f_n\| = 1$ and $\|Af_n\| \geq n^3$ for every n. Since f defined by

$$f = \sum_{n=1}^{\infty} \frac{f_n}{n^2}$$

is integrable and non-negative, we conclude that $f \geq f_n/n^2$, which implies that
$n^2 Af \geq Af_n$ for all $n \geq 1$. Thus we have for any n

$$n^3 \leq \|Af_n\| \leq n^2\|Af\|,$$

which is impossible, since $\|Af\| < \infty$.

4.1.2 Substochastic Semigroups

A family $\{P(t)\}_{t \geq 0}$ of linear operators on L^1 is called a *substochastic semigroup*
(*stochastic, positive semigroup*) if $P(t)$ is a substochastic (stochastic, positive) operator on L^1 for every t and $\{P(t)\}_{t \geq 0}$ is a strongly continuous semigroup. Thus a
substochastic semigroup is a positive contraction semigroup on L^1.

We proceed to characterize generators of substochastic semigroups using the result
of [83].

Theorem 4.1 (Phillips) *An operator $(A, \mathscr{D}(A))$ is the generator of a substochastic semigroup if and only if $\mathscr{D}(A)$ is dense in L^1, A is dispersive, i.e.*

$$\|(\lambda f - Af)^+\| \geq \lambda\|f^+\|, \quad f \in \mathscr{D}(A), \ \lambda > 0,$$

and the range of $\lambda I - A$ is L^1 for some $\lambda > 0$.

Let us outline the proof. Note that if A is dispersive and $f \in \mathscr{D}(A)$ then $-f \in \mathscr{D}(A)$ and, for any $\lambda > 0$,

$$\|(\lambda f - Af)^-\| = \|(\lambda(-f) - A(-f))^+\| \geq \lambda\|(-f)^+\| = \lambda\|f^-\|,$$

which shows that A is dissipative, since $\|f\| = \|f^+\| + \|f^-\|$ for any $f \in L^1$. Moreover, we have $\|(\lambda R(\lambda, A)f)^+\| \leq \|f^+\|$, which implies that the operator $\lambda R(\lambda, A)$ is substochastic for every $\lambda \in \rho(A)$, by Proposition 4.1. This shows that $(A, \mathscr{D}(A))$ is the generator of a substochastic semigroup. For the proof of the converse we refer the reader to [5, C-II].

We now characterize a dispersive operator using positive functionals. The dual space $(L^1)^*$ can be identified with $L^\infty = L^\infty(X, \Sigma, m)$, thus for each $\alpha \in (L^1)^*$ there exists a unique element $g \in L^\infty$ such that

$$\alpha(f) = \langle g, f \rangle = \int_X g(x)f(x)\, m(dx), \quad f \in L^1.$$

The operator A is dispersive if and only if for each $f \in \mathscr{D}(A)$ there exists $g \in L^\infty$, $g \geq 0$, with $\|g\|_\infty \leq 1$ such that

$$\langle g, f \rangle = \|f^+\| \quad \text{and} \quad \langle g, Af \rangle \leq 0.$$

In general, it is difficult to check that a given operator generates a substochastic semigroup using the Phillips theorem. We provide another approach in Sect. 4.1.4.

Remark 4.1 The Phillips theorem is valid in Banach lattices; see [5]. It can be used to generate a positive contraction semigroup on a closed subspace $\mathscr{X}$ of the space $B(X)$ of bounded measurable functions with the supremum norm $\|\cdot\|_u$. In particular, suppose that the operator $(A, \mathscr{D}(A))$ satisfies the *positive maximum principle* of the form: for each $f \in \mathscr{D}(A)$ such that $\|f^+\|_u > 0$ there exists $x_0 \in X$ satisfying $f(x_0) = \|f^+\|_u$ and $Af(x_0) \leq 0$. Then the operator A is dispersive, since

$$\|(\lambda f - Af)^+\|_u \geq \max\{0, \lambda f(x_0) - Af(x_0)\} \geq \lambda f(x_0) \geq \lambda\|f^+\|_u.$$

Remark 4.2 A uniformly continuous semigroup with generator A is a positive semigroup if and only if the operator $A + \lambda I$ is positive for some constant λ (see [5, C-II, Theorem 1.11]). Thus a substochastic (stochastic) uniformly continuous semigroup on $L^1(X, \Sigma, m)$ has the generator of the form $A = -\lambda I + \lambda P$, where λ is a non-negative constant and P is a substochastic (stochastic) operator (see [96, Section 3.2]).

Remark 4.3 Let $P: L^1(X, \Sigma, m) \to L^1(X, \Sigma, m)$ be a substochastic operator and $Y \in \Sigma$. The family of sets $\Sigma_Y = \{B \in \Sigma : B \subseteq Y\}$ is called the *restriction* of Σ to Y. By m_Y we denote the restriction of the measure m to Σ_Y. The *restriction* of P to the set Y is the substochastic operator $P_Y: L^1(Y, \Sigma_Y, m_Y) \to L^1(Y, \Sigma_Y, m_Y)$ defined by $P_Y f(x) = P\tilde{f}(x)$ for $x \in Y$, where $\tilde{f}(x) = f(x)$ for $x \in Y$ and $\tilde{f}(x) = 0$ for $x \notin Y$. Analogously, we define the *restriction of a substochastic semigroup* $\{P(t)\}_{t\geq 0}$ to a set Y as a substochastic semigroup $\{P_Y(t)\}_{t\geq 0}$ on $L^1(Y, \Sigma_Y, m_Y)$ given by $P_Y(t)f(x) = P(t)\tilde{f}(x)$ for $x \in Y$ and $t \geq 0$.

4.1.3 Resolvent Positive Operators

If $(A, \mathscr{D}(A))$ is the generator of a substochastic semigroup $\{S(t)\}_{t\geq 0}$ then it follows from (3.6) that for $\lambda > 0$ and $f \in L^1_+$ we have

$$R(\lambda, A)f = \int_0^\infty e^{-\lambda t} S(t) f \, dt \geq 0.$$

Thus the resolvent operator of A at each $\lambda > 0$ is positive. Following Arendt [4], we call a linear operator A *resolvent positive* if there exists $\omega \in \mathbb{R}$ such that $(\omega, \infty) \subseteq \rho(A)$ and $R(\lambda, A) \geq 0$ for all $\lambda > \omega$. The *spectral bound* of A is defined as

$$s(A) = \inf\{\omega \in \mathbb{R} : (\omega, \infty) \subseteq \rho(A) \text{ and } R(\lambda, A) \geq 0 \text{ for all } \lambda > \omega\}.$$

We need the following properties of resolvent positive operators.

Lemma 4.1 *Let A be a resolvent positive operator and let $s(A)$ be the spectral bound of A. Then for each $\mu > \lambda > s(A)$*

$$R(\lambda, A) \geq R(\mu, A) \geq 0.$$

Moreover, $s(A) \notin \rho(A)$ when $s(A) > -\infty$.

Proof By definition $s(A) < \infty$. The resolvent identity

$$R(\lambda, A) - R(\mu, A) = (\mu - \lambda)R(\mu, A)R(\lambda, A),$$

implies that $R(\lambda, A) \geq R(\mu, A)$ if $\mu > \lambda > s(A)$. Suppose now that $s(A) > -\infty$ and that $\mu = s(A) \in \rho(A)$. Then $R(\mu, A) \geq 0$ and for $\lambda < \mu$ such that $\|(\mu - \lambda)R(\mu, A)\| < 1$ we see, by Proposition 3.2, that $\lambda \in \rho(A)$ and

$$R(\lambda, A) = \sum_{n=0}^{\infty} (\mu - \lambda)^n R(\mu, A)^{n+1} \geq 0,$$

which gives a contradiction with the definition of $s(A)$.

We now give the following characterization [117] of positive perturbations of resolvent positive operators as resolvent positive operators.

Theorem 4.2 *Suppose that* $(A, \mathcal{D}(A))$ *is a resolvent positive operator and that* $(B, \mathcal{D}(B))$ *is a positive operator with* $\mathcal{D}(B) \supseteq \mathcal{D}(A)$. *The operator* $(A + B, \mathcal{D}(A))$ *is resolvent positive if and only if*

$$\lim_{n \to \infty} \|(BR(\lambda, A))^n\| = 0 \quad \text{for some } \lambda > s(A). \tag{4.1}$$

The following unpublished perturbation result [30] for generators of positive semigroups is only valid on an L^1 space. For its proof we refer the reader to the framework of Miyadera–Voigt perturbation theory (see [10] and [117]).

Theorem 4.3 (Desch) *Suppose that* $(A, \mathcal{D}(A))$ *is the generator of a positive semigroup on* L^1 *and that* B *with* $\mathcal{D}(B) \supseteq \mathcal{D}(A)$ *is a positive operator. If* $(A + B, \mathcal{D}(A))$ *is resolvent positive, then* $(A + B, \mathcal{D}(A))$ *is the generator of a positive semigroup.*

4.1.4 Generation Theorems

We now provide a generation theorem for substochastic semigroups on L^1. Recall that we denote by $\mathcal{D}(A)_+$ the set of all non-negative elements from the domain $\mathcal{D}(A)$ of an operator A.

Theorem 4.4 *A linear operator* $(A, \mathcal{D}(A))$ *is the generator of a substochastic semigroup on* L^1 *if and only if* $\mathcal{D}(A)$ *is dense in* L^1, *the operator* A *is resolvent positive, and*

$$\int_X Af(x) \, m(dx) \leq 0 \quad \text{for all } f \in \mathcal{D}(A)_+. \tag{4.2}$$

Proof Suppose first that $(A, \mathcal{D}(A))$ is the generator of a substochastic semigroup $\{S(t)\}_{t \geq 0}$. We know that $(A, \mathcal{D}(A))$ is resolvent positive. We also have for any $f \in \mathcal{D}(A)_+$

$$\int_X Af(x) \, m(dx) = \lim_{t \to \infty} \frac{1}{t} \int_X (S(t)f(x) - f(x)) \, m(dx)$$

$$= \lim_{t \to \infty} \frac{1}{t} (\|S(t)f\| - \|f\|) \leq 0.$$

To show that $(A, \mathcal{D}(A))$ is the generator we make use of the Hille–Yosida theorem. Since A is resolvent positive, A is closed and the resolvent $R(\lambda, A)$ is a positive operator for all $\lambda > s(A)$, by Lemma 4.1. Thus, it remains to check that $s(A) \leq 0$ and that $\lambda R(\lambda, A)$ is a contraction for all $\lambda > 0$. Take $f \in \mathscr{D}(A)_+$. Then, for any $\lambda > 0$, we have

$$\|(\lambda - A)f\| \geq \int_X (\lambda - A)f(x)\,m(dx) = \lambda\|f\| - \int_X Af(x)\,m(dx) \geq \lambda\|f\|.$$

Consequently, if $\lambda > \max\{0, s(A)\}$ then

$$\|g\| \geq \lambda\|R(\lambda, A)g\| \quad \text{for all } g \in L^1_+,$$

and $\lambda\|R(\lambda, A)\| \leq 1$. Suppose now that $s(A) > 0$. Let $\lambda_n > s(A)$ and $\lambda_n \downarrow s(A)$. We have

$$\|R(\lambda_n, A)\| \leq \frac{1}{\lambda_n} \leq \frac{1}{s(A)}, \quad n \geq 0.$$

Since $s(A) > -\infty$, we have $s(A) \notin \rho(A)$. Thus $\|R(\lambda_n, A)\| \to \infty$, which gives a contradiction and completes the proof.

Recall that an operator is stochastic if and only if it is substochastic and it preserves the L^1 norm on the positive cone. Thus Theorem 4.4 leads to the following

Corollary 4.1 *A linear operator $(A, \mathscr{D}(A))$ is the generator of a stochastic semigroup on L^1 if and only if $\mathscr{D}(A)$ is dense in L^1, the operator A is resolvent positive, and*

$$\int_X Af(x)\,m(dx) = 0 \quad \text{for all } f \in \mathscr{D}(A)_+.$$

Proof If $\{S(t)\}_{t \geq 0}$ is a stochastic semigroup, then $\|S(t)f\| = \|f\|$ for all $f \in L^1_+$. Hence

$$\int_X Af(x)\,m(dx) = \lim_{t \to \infty} \frac{1}{t}(\|S(t)f\| - \|f\|) = 0, \quad f \in \mathscr{D}(A)_+.$$

To show that a substochastic semigroup existing by Theorem 4.4 is stochastic let us take $f \in \mathscr{D}(A)_+$. Then $S(t)f \in \mathscr{D}(A)_+$ for $t > 0$ and

$$\frac{d}{dt}\|S(t)f\| = \frac{d}{dt}\int_X S(t)f(x)\,m(dx) = \int_X AS(t)f(x)\,m(dx) = 0.$$

4.1.5 Positive Perturbations

In this section, we provide sufficient conditions for the sum of two operators to be the generator of a substochastic semigroup. We assume that $(A, \mathscr{D}(A))$ is the generator of a substochastic semigroup and that $(B, \mathscr{D}(B))$ is an operator with $\mathscr{D}(B) \supseteq \mathscr{D}(A)$.

It follows from Theorem 4.4 that $(A + B, \mathscr{D}(A))$ is the generator of a substochastic semigroup if and only if $(A + B, \mathscr{D}(A))$ is resolvent positive and

$$\int_X (Af(x) + Bf(x)) \, m(dx) \leq 0 \quad \text{for} \quad f \in \mathscr{D}(A)_+. \tag{4.3}$$

We now suppose that $(A, \mathscr{D}(A))$ is resolvent positive and $(B, \mathscr{D}(A))$ is a positive operator. Then $R(\lambda, A)f \geq 0$ for every $\lambda > s(A)$ and $BR(\lambda, A)f \geq 0$ for $f \in L^1_+$ which implies that the positive operator $BR(\lambda, A)$ being defined everywhere is bounded. If, additionally, (4.3) holds then the operator $BR(\lambda, A)$ is substochastic for every $\lambda > 0$. To see this take $\lambda > 0$ and observe that

$$(\lambda I - A - B)R(\lambda, A)f = (I - BR(\lambda, A))f$$

for $f \in L^1$. Thus

$$BR(\lambda, A)f + \lambda R(\lambda, A)f = f + (A + B)R(\lambda, A)f \quad \text{for } f \in L^1_+. \tag{4.4}$$

This together with (4.3) implies that

$$\|BR(\lambda, A)f\| + \|\lambda R(\lambda, A)f\| \leq \|f\| \quad \text{for } f \in L^1_+, \tag{4.5}$$

showing that $\|BR(\lambda, A)\| \leq 1$. If there exists $\lambda > 0$ such that $\|BR(\lambda, A)\| < 1$ then $(A + B, \mathscr{D}(A))$ generates a substochastic semigroup, as we show next. The case when $\|BR(\lambda, A)\| = 1$ for all $\lambda > 0$ is discussed in the next section.

Theorem 4.5 *Let* $(A, \mathscr{D}(A))$ *be resolvent positive and* $(B, \mathscr{D}(A))$ *be a positive operator such that* (4.3) *holds. If* $\|BR(\lambda, A)\| < 1$ *for some* $\lambda > 0$ *then*

$$R(\lambda, A + B) = R(\lambda, A) \sum_{n=0}^{\infty} (BR(\lambda, A))^n$$

and the operator $(A + B, \mathscr{D}(A))$ *generates a substochastic semigroup.*

Proof Since $\|BR(\lambda, A)\| < 1$, the operator $I - BR(\lambda, A)$ is invertible. Its inverse is given by the Neumann expansion

$$(I - BR(\lambda, A))^{-1} = \sum_{n=0}^{\infty} (BR(\lambda, A))^n$$

and is a positive operator. The formula for $R(\lambda, A + B)$ follows from Lemma 3.3. Thus the operator $(A + B, \mathscr{D}(A))$ is resolvent positive, which together with Theorem 4.4 completes the proof.

We show that positive bounded perturbations of generators of substochastic semigroups lead to positive semigroups.

Corollary 4.2 *Suppose that* $(A, \mathcal{D}(A))$ *is the generator of a substochastic semi-group and that* B *with* $\mathcal{D}(B) = L^1$ *is a positive operator. Then* $(A + B, \mathcal{D}(A))$ *is the generator of a positive semigroup.*

Proof By rescaling, it is enough to show that the operator $(A + B - \|B\| I, \mathcal{D}(A))$ is the generator of a substochastic semigroup. Since B is bounded and $\lambda \|R(\lambda, A)\| \leq 1$, there exists $\lambda > 0$ such that $\|B\| < \lambda$, which implies that $\|BR(\lambda, A)\| < 1$. We have $R(\lambda, A) = R(\lambda - \|B\|, A - \|B\| I)$ and the operator $(A - \|B\| I, \mathcal{D}(A))$ generates a substochastic semigroup. For $f \in \mathcal{D}(A)_+$ we have

$$\int_X (A + B - \|B\| I) f(x)\, m(dx) = \int_X Af(x)\, m(dx) + \|Bf\| - \|B\| \|f\| \leq 0,$$

and the claim follows from Theorem 4.5.

We now consider positive perturbations of generators on L^1 with boundaries as studied in [45]. The case of unbounded positive operator B is treated in the next section.

Theorem 4.6 *Let* $L_0^1 = L^1(\Gamma_0, \Sigma_0, m_0)$, *where* $(\Gamma_0, \Sigma_0, m_0)$ *is a* σ-*finite measure space. We assume that* $(A, \mathcal{D})$ *is an operator on* L^1 *such that*

(1) for each $\lambda > 0$ *the operator* $\Psi_0 \colon \mathcal{D} \to L_0^1$ *restricted to the nullspace* $\mathcal{N}(\lambda I - A)$ *of the operator* $(\lambda I - A, \mathcal{D})$ *is invertible with positive inverse* $\Psi(\lambda) \colon L_0^1 \to \mathcal{N}(\lambda I - A)$;
(2) the operator $\Psi \colon \mathcal{D} \to L_0^1$ *is positive and there is* $c > 0$ *such that* $\|\lambda \Psi \Psi(\lambda)\| \leq c$ *for all* $\lambda > 0$;
(3) the restriction A_0 *of the operator* A *to* $\mathcal{D}(A_0) = \mathcal{N}(\Psi_0)$ *is resolvent positive.*

If B *is a bounded positive operator and condition* (4.3) *holds for* A *with* $\mathcal{D}(A) = \mathcal{N}(\Psi - \Psi_0)$, *then* $(A + B, \mathcal{D}(A))$ *is the generator of a substochastic semigroup.*

Proof Condition (2) implies that the operator $I - \Psi \Psi(\lambda)$ is invertible for $\lambda > c$. Since $(A_0, \mathcal{D}(A_0))$ is resolvent positive, the operators Ψ and $\Psi(\lambda)$ are positive and the operator $I - \Psi \Psi(\lambda)$ is invertible for all sufficiently large $\lambda > 0$, Lemma 3.4 implies that the operator $(A, \mathcal{D}(A))$ is resolvent positive. Consequently, we have $\|BR(\lambda, A)\| < 1$ for some $\lambda > 0$ and the result follows from Theorem 4.5.

We now show how we can generate stochastic semigroups by using perturbation theorems.

Theorem 4.7 *Suppose that* $(A_0, \mathcal{D}(A_0))$ *is the generator of a stochastic semigroup on* $L^1 = L^1(X, \Sigma, m)$ *and that* P *is a stochastic operator on* L^1. *Then for any bounded measurable function* $\varphi \colon X \to [0, \infty)$ *the operator*

$$A_1 f = A_0 f - \varphi f, \quad f \in \mathcal{D}(A_0),$$

is the generator of a substochastic semigroup and the operator

$$Af = A_1 f + P(\varphi f), \quad f \in \mathscr{D}(A_0),$$

is the generator of a stochastic semigroup.

Proof The function φ is bounded. Thus we can take $\lambda = \sup\{\varphi(x) : x \in X\}$. Then $\lambda - \varphi(x) \geq 0$ for all x. By rescaling, the operator $(A_0 - \lambda I, \mathscr{D}(A_0))$ is the generator of a substochastic semigroup. We have

$$A_1 f = A_0 f - \varphi f = A_0 f - \lambda f + (\lambda - \varphi) f \quad \text{for } f \in \mathscr{D}(A_0).$$

The operator $(A_0, \mathscr{D}(A_0))$ is the generator of a stochastic semigroup and $\varphi \geq 0$, hence

$$\int_X A_1 f(x) \, m(dx) = - \int_X \varphi(x) f(x) \, m(dx) \leq 0$$

for any $f \in \mathscr{D}(A_0)_+$. Since the operator $f \mapsto (\lambda - \varphi) f$ is positive and bounded, the operator $(A_1, \mathscr{D}(A_0))$ generates a substochastic semigroup, by Theorem 4.5. Finally, the operator $Bf = P(\varphi f)$ for $f \in L^1$ is bounded and positive. Since P preserves the integral on the cone L_+^1, we have

$$\int_X Bf(x) \, m(dx) = \int_X \varphi(x) f(x) \, m(dx), \quad f \in L_+^1.$$

Consequently, the operator $(A_1 + B, \mathscr{D}(A_0))$ is resolvent positive and is the generator of a stochastic semigroup, by Corollary 4.1. □

Remark 4.4 Note that if A_0 generates a substochastic semigroup and P is a substochastic operator then A as defined in Theorem 4.7 generates a substochastic semigroup.

4.1.6 Positive Unbounded Perturbations

In this section we describe a generation result of Kato [56] (see also [9, 10, 116]) ensuring that some extension of $(A + B, \mathscr{D}(A))$ is the generator of a substochastic semigroup when the operator B is positive and unbounded.

Theorem 4.8 *Assume that $(A, \mathscr{D}(A))$ is the generator of a substochastic semigroup on L^1 and that $(B, \mathscr{D}(A))$ is a positive operator such that (4.3) holds. There exists an extension $(C, \mathscr{D}(C))$ of the operator $(A + B, \mathscr{D}(A))$ which is the generator of a substochastic semigroup $\{P(t)\}_{t \geq 0}$. The resolvent operator of C is*

$$R(\lambda, C)f = \lim_{N \to \infty} R(\lambda, A) \sum_{n=0}^{N} (BR(\lambda, A))^n f, \quad f \in L^1, \ \lambda > 0. \qquad (4.6)$$

The semigroup $\{P(t)\}_{t\geq 0}$ from Theorem 4.8 is called the *minimal semigroup* related to $A + B$. To justify the name we show that the semigroup $\{P(t)\}_{t\geq 0}$ is the smallest substochastic semigroup whose generator is an extension of $(A + B, \mathscr{D}(A))$, i.e. for any positive semigroup $\{\widetilde{P}(t)\}_{t\geq 0}$ with generator $(\widetilde{C}, \mathscr{D}(\widetilde{C}))$ being an extension of the operator $(A + B, \mathscr{D}(A))$ we have $\widetilde{P}(t)f \geq P(t)f$ for all $f \in L^1_+$, $t > 0$. To this end for each $r \in (0, 1)$ consider the operator $A_r = A + rB$ with domain $\mathscr{D}(A)$. We have $\|rBR(\lambda, A)\| \leq r < 1$ and the operator $(A_r, \mathscr{D}(A))$ is resolvent positive with

$$R(\lambda, A + rB)f = R(\lambda, A)\sum_{n=0}^{\infty}(rBR(\lambda, A))^n f, \quad f \in L^1, \ \lambda > 0.$$

For $f \in \mathscr{D}(A)$ we have

$$(\lambda I - A - rB)f - (\lambda I - \widetilde{C})f = (1 - r)Bf$$

which leads to

$$R(\lambda, \widetilde{C}) - R(\lambda, A + rB) = (1 - r)R(\lambda, \widetilde{C})BR(\lambda, A + rB) \geq 0$$

for $\lambda > s(\widetilde{C})$. Hence, for each $f \in L^1_+$ we have

$$R(\lambda, \widetilde{C})f \geq R(\lambda, A + rB)f \geq R(\lambda, A)\sum_{n=0}^{N}(rBR(\lambda, A))^n f$$

for all $r \in (0, 1)$ and $N \geq 0$. Taking the limit as $r \to 1$ and then as $N \to \infty$, we obtain

$$R(\lambda, \widetilde{C})f \geq R(\lambda, C)f, \quad f \in L^1_+,$$

which implies that $\widetilde{P}(t)f \geq P(t)f$, by (3.7).

The minimal semigroup $\{P(t)\}_{t\geq 0}$ can be defined by

$$P(t)f = \lim_{r \to 1} P_r(t)f, \quad f \in L^1, \ t > 0,$$

where, for each $r \in (0, 1)$, $\{P_r(t)\}_{t\geq 0}$ is a substochastic semigroup generated by $(A + rB, \mathscr{D}(A))$. The semigroup $\{P(t)\}_{t\geq 0}$ satisfies the integral Eq. (3.12) and it is also given by the Dyson–Phillips expansion (3.13), where $\{S(t)\}_{t\geq 0}$ is the semigroup generated by $(A, \mathscr{D}(A))$. Theorem 4.8 and the Lumer–Phillips theorem imply that the closure of the operator $(A + B, \mathscr{D}(A))$ is the generator of a substochastic semigroup if and only if the range of $\lambda I - (A + B)$ is dense in L^1 for some $\lambda > 0$. Equivalently, by Corollary 3.1, the operator $\lambda I^* - (A + B)^*$ is one to one. By Theorem 3.6, what we are missing is the characterisation of a core of the generator of the minimal semigroup. Necessary and sufficient conditions for the generator of the minimal semigroup $\{P(t)\}_{t \geq 0}$ to be the closure of the operator $(A + B, \mathcal{D}(A))$ are to be found in [10, 113, 114]. In particular, we have the following.

Theorem 4.9 *Let* $\lambda > 0$ *and let* $\{P(t)\}_{t \geq 0}$ *be the minimal semigroup related to* $A + B$ *satisfying*

$$\int_X (Af(x) + Bf(x)) \, m(dx) = 0 \quad \text{for} \quad f \in \mathcal{D}(A)_+. \tag{4.7}$$

Then the following are equivalent

(1) *The minimal semigroup* $\{P(t)\}_{t \geq 0}$ *related to* $A + B$ *is stochastic.*
(2) *For all* $f \in L^1$

$$\lim_{n \to \infty} \|(BR(\lambda, A))^n f\| = 0. \tag{4.8}$$

(3) *If for some* $g \in L^\infty$, $g \geq 0$, *we have* $(BR(\lambda, A))^* g = g$ *then* $g = 0$, *where* $(BR(\lambda, A))^*$ *denotes the adjoint of* $BR(\lambda, A)$.
(4) *The closure of* $(A + B, \mathcal{D}(A))$ *is the generator of a substochastic semigroup.*

Remark 4.5 Observe that condition (4.8) is weaker than condition (4.1), the latter implying that the operator $A + B$ is resolvent positive which together with (4.7) implies that $(A + B, \mathcal{D}(A))$ is the generator. In particular, it is shown in [114] that under assumption (4.3) the operator $(A + B, \mathcal{D}(A))$ is the generator of a substochastic semigroup if and only if the operator $BR(\lambda, A)$ is *quasi-compact* for some $\lambda > 0$, i.e. there exists a linear compact operator K and $n \in \mathbb{N}$ such that $\|(BR(\lambda, A))^n - K\| < 1$. Recall that an operator K is called *compact* if the closure of the image of any bounded set under the mapping K is a compact subset of L^1.

Remark 4.6 If A is the generator of a substochastic semigroup then $R(\lambda, A)^* = R(\lambda, A^*)$ for all $\lambda > 0$. Moreover, we have

$$(BR(\lambda, A))^* g = R(\lambda, A^*)B^* g \quad \text{for } g \in \mathcal{D}(B^*).$$

Hence, any $g \in L^\infty$, $g \geq 0$, is a solution of $(BR(\lambda, A))^* g = g$ if and only if $g \in \mathcal{D}(A^*) \cap \mathcal{D}(B^*)$ and

$$A^* g + B^* g = \lambda g.$$

Note that if we let

$$Bf = P(\varphi f) \quad \text{for } f \in L^1_\varphi = \{f \in L^1 : \varphi f \in L^1\},$$

then $\mathcal{D}(B^*) = \{g \in L^\infty : \varphi P^* g \in L^\infty\}$. If, additionally, $Af = -\varphi f$, then $\mathcal{D}(A^*) = \{g \in L^\infty : \varphi g \in L^\infty\}$) and the adjoint of the operator $BR(\lambda, A)$ is given by

$$(BR(\lambda, A))^* g = \frac{\varphi}{\lambda + \varphi} P^* g.$$

4.1.7 Adjoint and Transition Semigroups

We say that a substochastic semigroup $\{P(t)\}_{t\geq 0}$ *corresponds to a transition function*
$P = \{P(t, \cdot): t \geq 0\}$ if for each $t > 0$ the adjoint operator $P^*(t)$ of $P(t)$ is given by

$$P^*(t)g(x) = \int_X g(y)P(t, x, dy), \quad g \in L^\infty.$$

Equivalently, each transition kernel $P(t, \cdot)$ satisfies (2.4), or

$$\int_B P(t)f(x)\, m(dx) = \int_X P(t, x, B)f(x)\, m(dx), \quad B \in \Sigma, \ f \in L^1_+. \tag{4.9}$$

In particular, if a stochastic semigroup $\{P(t)\}_{t\geq 0}$ corresponds to the transition function P induced by a Markov process $\xi = \{\xi(t): t \geq 0\}$, i.e.

$$P(t, x, B) = \mathbb{P}_x(\xi(t) \in B), \quad x \in X, \ B \in \Sigma,$$

where $\mathbb{P}_x$ is the distribution of $\xi(t)$ starting at x, then $P(t)f$ is the density of $\xi(t)$ if the distribution of $\xi(0)$ has a density f.

In this section, we relate a stochastic semigroup $\{P(t)\}_{t\geq 0}$ to the transition semigroup $\{T(t)\}_{t\geq 0}$ on $B(X)$ associated to a homogeneous Markov process $\xi = \{\xi(t): t \geq 0\}$ with transition function $P = \{P(t, \cdot): t \geq 0\}$ as described in Sect. 2.3.4. The transition semigroup $\{T(t)\}_{t\geq 0}$ on the Banach space $B(X)$ with supremum norm $\|\cdot\|_u$ associated to the process ξ is given by

$$T(t)g(x) = \mathbb{E}_x(g(\xi(t))) = \int_X g(y)P(t, x, dy), \quad g \in B(X).$$

In particular, if

$$\int_X T(t)g(x)f(x)\, m(dx) = \int_X g(x)P(t)f(x)\, m(dx) \tag{4.10}$$

holds for all $f \in L^1_+$ and all $g \in B(X)$, then (4.9) holds and $\{P(t)\}_{t\geq 0}$ corresponds to the transition function P.

The transition semigroup $\{T(t)\}_{t\geq 0}$ on $B(X)$ is strongly continuous on the closed subspace of $B(X)$

$$B_0(X) = \{g \in B(X): \lim_{t\downarrow 0} \|T(t)g - g\|_u = 0\}$$

and $T(t)g \in B_0(X)$ for $g \in B_0(X)$. The generator $(L, \mathscr{D}(L))$ of the semigroup $\{T(t)\}_{t\geq 0}$ is densely defined in $B_0(X)$. On the other hand for each t, the adjoint $P^*(t)$ of the operator $P(t)$ is a contraction on L^∞

$$\int_X P^*(t)g(x)f(x)\,m(dx) = \int_X g(x)P(t)f(x)\,m(dx), \quad g \in L^\infty, \; f \in L^1,$$

and $\{P^*(t)\}_{t\geq 0}$ is a semigroup on L^∞, called the *adjoint semigroup*. We would like to relate the semigroups by using their generators. We define the subspace of strong continuity of the adjoint semigroup $\{P^*(t)\}_{t\geq 0}$ by

$$L^\odot = \{g \in L^\infty : \lim_{t\downarrow 0} \|P^*(t)g - g\|_\infty = 0\},$$

where $\|\cdot\|_\infty$ denotes the essential supremum norm in L^∞. It is called the *sun dual* (or the semigroup dual) space. The semigroup given by the part of $P^*(t)$ in $L^\odot$, i.e.

$$P^\odot(t)g = P^*(t)g \quad \text{for } g \in L^\odot,$$

is called the *sun dual semigroup*. It is strongly continuous, thus a contraction semigroup on $L^\odot$. Let $(A, \mathscr{D}(A))$ be the generator of the semigroup $\{P(t)\}_{t\geq 0}$ and let $(A^*, \mathscr{D}(A^*))$ be its adjoint. It is easy to see that $\mathscr{D}(A^*) \subset L^\odot$ and that A^* is an extension of the generator $A^\odot$ of the sun dual semigroup $\{P^\odot(t)\}_{t\geq 0}$. In fact, $(A^\odot, \mathscr{D}(A^\odot))$ is the part of $(A^*, \mathscr{D}(A^*))$ in $L^\odot$, i.e.

$$A^\odot g = A^* g \quad \text{for } g \in \mathscr{D}(A^\odot) = \{g \in \mathscr{D}(A^*) : A^* g \in L^\odot\}.$$

Consequently, if Eq. (4.10) holds for all $g \in B_0(X)$ and $f \in L^1$ then $\{T(t)\}_{t\geq 0}$ is the part of the sun dual semigroup $\{P^\odot(t)\}_{t\geq 0}$ in $B_0(X)$ and we have $L \subseteq A^\odot \subseteq A^*$.

We would like to obtain (4.10) using generators. We now assume that there exists a closed subspace $\mathscr{X}$ of $B_0(X)$ such that $T(t)g \in \mathscr{X}$ for $g \in \mathscr{X}$ and that $\mathscr{D}_L$ is a core for the part of the generator $(L, \mathscr{D}(L))$ of $\{T(t)\}_{t\geq 0}$ in $\mathscr{X}$. We suppose that the following condition holds

$$\int_X Lg(x)f(x)\,m(dx) = \int_X g(x)Af(x)\,m(dx) \tag{4.11}$$

for $g \in \mathscr{D}_L$ and $f \in \mathscr{D}_A$, where $\mathscr{D}_A$ is a core for the generator $(A, \mathscr{D}(A))$ of the stochastic semigroup $\{P(t)\}_{t\geq 0}$ on L^1. By approximations, condition (4.11) extends to all functions from the domains of the generators. It is easy to see that the function

$$v(s) = \int_X T(s)g(x)P(t-s)f(x)\,m(dx), \quad s \in [0, t], \; g \in \mathscr{D}(L), \; f \in \mathscr{D}(A),$$

is continuous on $[0, t]$ and differentiable with $v'(s) = 0$ for $s \in (0, t)$. This implies that (4.10) holds for $g \in \mathscr{D}(L)$ and $f \in \mathscr{D}(A)$. Since domains of the generators are dense in the corresponding spaces, we conclude that $P^*(t)g = T(t)g$ for $g \in \mathscr{X}$.

If $\mathscr{X}$ is large enough so that we can extend equality (4.10) to all g being indicator functions, then $\{P(t)\}_{t\geq 0}$ corresponds to the transition function P. In particular, if X is a metric space and $\Sigma = \mathscr{B}(X)$, then, by the uniqueness theorem A.2, if condition

(4.9) holds for all closed subsets B, then it holds for all Borel sets. Since each indicator function of a closed set can be approximated by a sequence of (globally) Lipschitz continuous functions, it is enough to check (4.10) for uniformly continuous functions $g\colon X \to \mathbb{R}$, or only for Lipschitz continuous functions.

To check condition (4.11) we can use the extended generator $(\widetilde{L}, \mathscr{D}(\widetilde{L}))$, which for a class of PDMPs is characterized at the end of Sect. 2.3.6. The next result may be useful in showing that L is a generator. The notion of resolvent positive operators as introduced in Sect. 4.1.3 can also be used for operators defined on $\mathscr{X} \subset B(X)$ with the supremum norm.

Theorem 4.10 *Let $(L, \mathscr{D}(L))$ be a resolvent positive operator in a closed subspace $\mathscr{X}$ of $B(X)$. If $\mathbf{1}_X \in \mathscr{D}(L)$ then L is the generator of a positive semigroup on the closure of $\mathscr{D}(L)$. If, additionally, $L\mathbf{1}_X = 0$ then L generates a contraction semigroup.*

Proof The assumption $\mathbf{1}_X \in \mathscr{D}(L)$ implies that the operator L is cofinal, i.e. for each non-negative $g \in \mathscr{X}$ there exists a non-negative $h \in \mathscr{D}(L)$ such that $g(x) \le h(x)$ for all $x \in X$. As in the proof of [4, Theorem 2.1] we conclude that L is a Hille–Yosida operator. Consequently, L generates a positive semigroup on the closure of $\mathscr{D}(L)$ (see Remark 3.5). $\qquad\qquad\blacksquare$

However, in general it is difficult to determine a sufficiently large space $\mathscr{X}$. Thus, in the next section we will also prove directly condition (4.10) by either using the formula for the transition function or by using the Dyson–Phillips expansion for the semigroup $\{P(t)\}_{t\ge 0}$ and the Kolmogorov equation (2.31).

4.2 Stochastic Semigroups for PDMPs

4.2.1 *Jump-Type Markov Processes*

In Sect. 1.3 we considered a pure jump-type Markov process. This is a Markov process on the space (X, Σ, m) which remains constant between jumps; see Sect. 2.3.3. In the case when the state space X is at most a countable set, we obtain continuous-time Markov chains. We assume that a measurable function $\varphi\colon X \to [0, \infty)$ is a jump rate function. Let a transition probability function $P(x, B)$ describe the new position of the point after a jump from x. We assume that $P(x, B)$ satisfies (2.4) and that P is the stochastic operator on $L^1(X, \Sigma, m)$ corresponding to $P(x, B)$.

Suppose first that φ is bounded. We now show that if the random variable $\xi(t)$ is the position of the moving point at time t and $\xi(0)$ has density $u(0)$, then $\xi(t)$ has density $u(t)$ which satisfies the evolution equation

$$u'(t) = Au(t) = -\varphi u(t) + P(\varphi u(t)). \tag{4.12}$$

To this end we let $\lambda = \sup\{\varphi(x)\colon x \in X\}$ and

$$\bar{P}f(x) = \lambda^{-1}\big(P(\varphi f)(x) + (\lambda - \varphi(x))f(x)\big).$$

Then $\bar{P}$ is a stochastic operator and $A = -\lambda I + \lambda\bar{P}$. The stochastic operator $\bar{P}$ corresponds to the transition probability $\bar{P}$ defined in (2.20). It follows from Corollary 3.2 that the solution of (4.12) is given by

$$u(t) = \sum_{k=0}^{\infty} \frac{(\lambda t)^k e^{-\lambda t}}{k!} \bar{P}^k u(0). \tag{4.13}$$

The stochastic semigroup $\{P(t)\}_{t\geq 0}$ defined by $P(t)f = u(t)$ with $f = u(0)$ corresponds to the transition function $P(t, x, B)$, induced by the process $\xi = \{\xi(t): t \geq 0\}$ (see Sect. 4.1.7) and given by (2.21), since

$$\int_B \bar{P}^k f(x)\, m(dx) = \int_X \bar{P}^k(x, B) f(x)\, m(dx)$$

for all k, $f \in L^1_+$ and $B \in \Sigma$.

Suppose now that φ is unbounded and that

$$Af = -\varphi f + P(\varphi f) \quad \text{for } f \in L^1_\varphi = \{f \in L^1: \varphi f \in L^1\}.$$

The operator $L^1_\varphi \ni f \mapsto -\varphi f$ is the generator of a substochastic semigroup, by Sect. 3.2.2, and the operator $L^1_\varphi \ni f \mapsto P(\varphi f)$ is positive. For $f \in L^1_\varphi$ and $f \geq 0$, we have

$$\int_X Af(x)\, m(dx) = \int_X -\varphi(x) f(x)\, m(dx) + \int_X P(\varphi f)(x)\, m(dx) = 0,$$

since P preserves the integral of non-negative integrable elements. Theorem 4.8 implies that there exists the minimal substochastic semigroup $\{P(t)\}_{t\geq 0}$ generated by an extension of the operator (A, L^1_φ). From Theorem 4.9 and Remark 4.6 it follows that the minimal semigroup related to A is stochastic if and only if every non-negative and essentially bounded solution g of the equation

$$\varphi(P^*g - g) = \lambda g$$

is equal to 0 a.e., where P^* is the adjoint of P and $\lambda > 0$.

We close this section by providing the Kato original result [56, Theorem 3] in the setting of $l^1 = L^1(I, \Sigma, m)$ where I is at most a countable set, Σ is the σ-algebra of all subsets of I, and m is the counting measure on I. Let $Q = [q_{ij}]_{i,j\in I}$ be a matrix, or an infinite dimensional matrix, when I is infinite. We consider a linear operator (bounded or unbounded) on l^1 given by $(Qu)_i = \sum_{j\in I} q_{ij} u_j$ for $i \in I$. The matrix $Q = [q_{ij}]_{i,j\in I}$ is called a *Kolmogorov matrix*, if its entries have the following properties

(i) $q_{ij} \geq 0$ for $i \neq j$,

(ii) $\sum_{i \in I} q_{ij} = 0$ for $j \in I$,

and it is called a *sub-Kolmogorov matrix* if it satisfies condition (i) and the condition

(ii′) $\sum_{i \in I} q_{ij} \leq 0$ for $j \in I$.

Let us define the jump rate function $\varphi: I \to [0, \infty)$ by $\varphi = (\varphi_j)_{j \in I}$ with $\varphi_j = -q_{jj}$, and the jump transition kernel by $P(j, \{i\}) = p_{ij}$ with $p_{ii} = 0$ and

$$p_{ij} = \frac{q_{ij}}{\varphi_j} \ \text{ if } \varphi_j > 0 \ \text{ and } \ p_{ij} = 1 \ \text{ if } \varphi_j = 0, \ \ i \neq j.$$

The transition kernel P defines the substochastic operator $P: l^1 \to l^1$ by

$$(Pu)_i = \sum_{j \in I} p_{ij} u_j, \quad i \in I, u \in l^1.$$

We have

$$(Qu)_i = q_{ii} u_i + \sum_{j \neq i} q_{ij} u_j = -\varphi_i u_i + P(\varphi u)_i.$$

If Q is a sub-Kolmogorov matrix then condition (4.3) holds and the minimal substochastic semigroup $\{P(t)\}_{t \geq 0}$ related to Q is generated by an extension of the operator Q. Note that the matrix Q is a Kolmogorov matrix if and only if

$$\varphi_j = \sum_{i \neq j} q_{ij}, \quad j \in I,$$

or equivalently P is a stochastic operator. The matrix Q is a bounded operator on l^1 if and only if φ is bounded:

$$\sup_j |q_{jj}| < \infty.$$

In this case the solution of (4.12) is given by (4.13). If φ is unbounded then we have the following result.

Theorem 4.11 (Kato) *Let Q be a Kolmogorov matrix and let $\lambda > 0$ be a positive constant. We denote $Q^* = (q_{ij}^*)_{i, j \in I}$, where $q_{ij}^* = q_{ji}$ for $i, j \in I$. The minimal semigroup related to Q is a stochastic semigroup on l^1 if and only if the equation $Q^* w = \lambda w$ has no nonzero solution $w \in l^\infty$ and $w \geq 0$.*

A Kolmogorov matrix Q is called *non-explosive* if the minimal semigroup related to Q is stochastic.

4.2.2 Semigroups for Semiflows

Let (X, Σ, m) be a σ-finite measure space. Consider a measurable semiflow on X, i.e. $\pi : [0, \infty) \times X \to X$ is such that the mapping $(t, x) \mapsto \pi(t, x)$ is measurable and the transformations $\pi_t : X \to X$, where $\pi_t(x) = \pi(t, x)$ for $x \in X$ and $t \geq 0$, satisfy the semigroup property

$$\pi_0(x) = x \quad \text{and} \quad \pi_{s+t}(x) = \pi_t(\pi_s(x)), \quad t, s \geq 0, \ x \in X.$$

We assume that for each $t > 0$ the transformation π_t is non-singular with respect to m. Let $P_0(t)$ be the Frobenius–Perron operator on L^1 corresponding to π_t (see Sect. 2.1.5). It satisfies

$$\int_X \mathbf{1}_B(\pi_t(x)) f(x) m(dx) = \int_B P_0(t) f(x) m(dx), \quad B \in \Sigma, \ f \in L^1,$$

which, by linearity and approximations, can be extended to

$$\int_X g(\pi_t(x)) f(x) m(dx) = \int_X g(x) P_0(t) f(x) m(dx) \tag{4.14}$$

for all measurable and non-negative g, as well as for all $g \in L^\infty$ and $f \in L^1$. Since $\{\pi_t\}_{t \geq 0}$ is a semigroup on X, it follows from (4.14) that $P_0(0) = I$ and $P_0(t + s) = P_0(t)P_0(s)$ for all $s, t \geq 0$. Thus the family $\{P_0(t)\}_{t \geq 0}$ is a semigroup on L^1.

We give sufficient conditions for the semigroup of Frobenius–Perron operators to be a stochastic semigroup. The proof of the following is based on [45].

Theorem 4.12 *Assume that L^1 is separable. If the semigroup $\{P_0(t)\}_{t \geq 0}$ of Frobenius–Perron operators corresponding to $\{\pi_t\}_{t \geq 0}$ satisfies $P_0(t)|f| = |P_0(t)f|$ for all $f \in L^1$ and $t > 0$, then $\{P_0(t)\}_{t \geq 0}$ is a stochastic semigroup.*

Proof We first prove that for any $f \in L^1$ the function $t \mapsto P_0(t)f$ is continuous at each $s > 0$. If $g \in L^\infty_+$ and $f \in L^1_+$ then the function $(t, x) \mapsto g(\pi(t, x))f(x)$ is measurable and non-negative, thus the function

$$t \mapsto \int_X g(\pi(t, x)) f(x) m(dx)$$

is Borel measurable, by Fubini's theorem. This together with (4.14) implies that for any $f \in L^1$ the function

$$t \mapsto \langle g, P_0(t)f \rangle = \int_X g(x) P_0(t) f(x) m(dx)$$

is Borel measurable for each $g \in L^\infty$. Consequently, the function $t \mapsto P_0(t)f$ is weakly measurable. The space L^1 is separable. By Dunford's theorem [33], the function $t \mapsto P_0(t)f$ is continuous at $s > 0$. Next, we prove that

$$\|P_0(t)f - f\| = \|P_0(t+s)f - P_0(s)f\|$$

for any $t, s > 0$ and $f \in L^1$. Since $\pi(s, x) \in X$ for all $x \in X$ and $s > 0$, we have

$$\|P_0(t)f - f\| = \int_X 1_X(\pi(s, x))|P_0(t)f(x) - f(x)|\, m(dx),$$

and, by (4.14), we obtain

$$\|P_0(t)f - f\| = \int_X P_0(s)|P_0(t)f(x) - f(x)|\, m(dx).$$

By assumption $P_0(s)|P_0(t)f - f| = |P_0(t+s)f - P_0(s)f|$, which gives the claim and implies that $t \mapsto P_0(t)f$ is continuous at 0 for all $f \in L^1$.

We now assume that X is a separable metric space and that $\Sigma = \mathcal{B}(X)$ is the σ-algebra of Borel subsets of X. In that case the space $L^1 = L^1(X, \mathcal{B}(X), m)$ is separable for any σ-finite measure m on $(X, \mathcal{B}(X))$. If $\pi : [0, \infty) \times X \to X$ is a semiflow as in Sect. 1.19, then the mapping $(t, x) \mapsto \pi(t, x)$ is measurable. Moreover, if π_t is non-singular and one to one then the Frobenius–Perron operator $P_0(t)$ for the transformation π_t is of the form given in Sect. 2.1.5 and satisfies the condition $|P_0(t)f| = P_0(t)|f|$ for $f \in L^1$.

Recall that the survival function $t \mapsto \Phi_x(t)$ is right-continuous with $\Phi_x(0) = 1$ and that for each $t \geq 0$ the mapping $x \mapsto \Phi_x(t)$ is measurable. Define the operators $S(t), t \geq 0$, on L^1 by

$$S(t)f = P_0(t)(\Phi(t)f), \tag{4.15}$$

for $t \geq 0$, $f \in L^1$, where we write $\Phi(t)$ for the function $x \mapsto \Phi_x(t)$. From (4.14) it follows that

$$\int_X \Phi_x(t)1_B(\pi_t(x))f(x)\, m(dx) = \int_B S(t)f(x)\, m(dx) \tag{4.16}$$

for all $t \geq 0$, $f \in L^1_+$, $B \in \Sigma$.

Theorem 4.13 *Suppose that $\{S(t)\}_{t \geq 0}$ is defined as in (4.15) and satisfies (4.16). If $\{P_0(t)\}_{t \geq 0}$ is a substochastic semigroup then $\{S(t)\}_{t \geq 0}$ is a substochastic semigroup.*

Proof Since $0 \leq \Phi(t) \leq 1$, we obtain

$$0 \leq S(t)f \leq P_0(t)f$$

for $f \in L^1_+$. Thus, each operator $S(t)$ is substochastic. To show the semigroup property, we use the equation

$$\Phi_x(t+s) = \Phi_{\pi_s(x)}(t)\Phi_x(s), \quad x \in X, t, s \geq 0.$$

This and (4.14) imply that for any measurable set B and any $f \in L^1_+$ we have

$$\int_B P_0(s)(\Phi(t+s)f)(x)\,m(dx) = \int_X 1_B(\pi_s(x))\Phi_{\pi_s(x)}(t)\Phi_x(s)f(x)\,m(dx)$$

$$= \int_X 1_B(x)\Phi_x(t)P_0(s)(\Phi(s)f)(x)\,m(dx).$$

Consequently, $P_0(s)(\Phi(t+s)f) = \Phi(t)P_0(s)(\Phi(s)f)$ for $s, t \geq 0$, $f \in L^1_+$, which shows that

$$P_0(t+s)(\Phi(t+s)f) = P_0(t)(\Phi(t)P_0(s)(\Phi(s)f)) = S(t)(S(s)f).$$

Finally, observe that

$$\|S(t)f - f\| \leq \|\Phi(t)f - f\| + \|P_0(t)f - f\|$$

and that $\|\Phi(t)f - f\| \to 0$ as $t \downarrow 0$, which implies the strong continuity of the semigroup $\{S(t)\}_{t\geq 0}$.

Remark 4.7 We can use results of this section for the semiflow $\pi : [0, \infty) \times \tilde{X} \to \tilde{X}$ as in Sect. 1.19 with $\tilde{X}$ being an extension of the state space X. If the assumptions of Theorem 4.12 hold for the semigroup $\{\tilde{P}_0(t)\}_{t\geq 0}$ satisfying

$$\int_{\tilde{X}} g(\pi_t(x))f(x)\,\tilde{m}(dx) = \int_{\tilde{X}} g(x)\tilde{P}_0(t)f(x)\,\tilde{m}(dx)$$

for $f \in L^1(\tilde{X}, \tilde{\Sigma}, \tilde{m})$ and $g \in L^\infty(\tilde{X}, \tilde{\Sigma}, \tilde{m})$, then the restriction of $\{\tilde{P}_0(t)\}_{t\geq 0}$ to X is a substochastic semigroup $\{P_0(t)\}_{t\geq 0}$ on $L^1 = L^1(X, \Sigma, m)$. In particular, we have

$$\int_X 1_{[0,t_*(x))}(t)g(\pi_t(x))f(x)\,m(dx) = \int_X g(x)P_0(t)f(x)\,m(dx), \quad f \in L^1, \ g \in L^\infty,$$

where $t_*(x)$ is the exit time from the set X as defined in (1.39). If the survival function Φ is given by (1.41) then the semigroup $\{S(t)\}_{t\geq 0}$ satisfies (4.16).

4.2.3   PDMPs Without Boundaries

In this section we show how we can define stochastic semigroups for a large class of PDMPs. Let (X, Σ, m) be a σ-finite measure space, where X is a separable metric space and $\Sigma = \mathscr{B}(X)$ is the σ-algebra of Borel subsets of X. The characteristics (π, Φ, P) define a PDMP process $\xi = \{\xi(t): t \geq 0\}$ with state space X, where

$\pi : [0, \infty) \times X \to X$ is a semiflow as in Sect. 1.19, Φ is a survival function, and P is a jump distribution. The semiflow is measurable. To simplify the presentation we now assume that the active boundary is empty and also that the survival function Φ is defined by

$$\Phi_x(t) = \exp \left\{ - \int_0^t \varphi(\pi_r(x)) \, dr \right\}, \quad x \in X, \, t \geq 0, \tag{4.17}$$

where φ is a bounded measurable function such that for each $x \in X$, $\varphi(\pi_t(x)) \to \varphi(x)$ as $t \to 0$.

Theorem 4.14 *Let $\{P_0(t)\}_{t \geq 0}$ be a stochastic semigroup of Frobenius–Perron operators corresponding to $\{\pi_t\}_{t \geq 0}$. Suppose that $\{S(t)\}_{t \geq 0}$ is a substochastic semigroup as defined in (4.15) with $\Phi_x(t)$ as in (4.17). Then the generator of $\{S(t)\}_{t \geq 0}$ is given by*

$$A_1 f = A_0 f - \varphi f \quad \text{for } f \in \mathcal{D}(A_0), \tag{4.18}$$

where $(A_0, \mathcal{D}(A_0))$ is the generator of $\{P_0(t)\}_{t \geq 0}$.

Proof It follows from Theorem 4.7 that the operator $(A_1, \mathcal{D}(A_0))$ is the generator of a substochastic semigroup. By Proposition 3.1, it is enough to show that the generator of the semigroup $\{S(t)\}_{t \geq 0}$ is an extension of $(A_1, \mathcal{D}(A_0))$. Take $f \in \mathcal{D}(A_0)$ and observe that

$$\left\| \frac{1}{t}(S(t)f - f) - A_1 f \right\| \leq \left\| \frac{1}{t}(P_0(t)f - f) - A_0 f \right\| + \|\varphi f - P_0(t)(\varphi f)\|$$
$$+ \left\| P_0(t) \left(\frac{1}{t}(\Phi(t) - 1)f + \varphi f \right) \right\|.$$

We have

$$\lim_{t \downarrow 0} \frac{1}{t}(\Phi_x(t) - 1) = -\varphi(x), \quad x \in X,$$

which implies, by the dominated convergence theorem, that

$$\lim_{t \downarrow 0} \left\| \frac{1}{t}(\Phi(t) - 1)f + \varphi f \right\| = 0.$$

Since $\{P_0(t)\}_{t \geq 0}$ is a stochastic semigroup with generator A_0, we conclude that the generator of the semigroup $\{S(t)\}_{t \geq 0}$ is an extension of the operator $(A_1, \mathcal{D}(A_0))$.

Let the transition probability P correspond to a stochastic operator $P(x, B)$ on $L^1 = L^1(X, \Sigma, m)$. It follows from Theorem 4.7 that the operator

$$Af = A_0 f - \varphi f + P(\varphi f), \quad f \in \mathcal{D}(A_0),$$

is the generator of a stochastic semigroup $\{P(t)\}_{t \geq 0}$. According to the Dyson–Phillips formula (3.13), we have

$$P(t)f = \sum_{n=0}^{\infty} S_n(t)f, \quad \text{where} \quad S_n(t)f = \int_0^t S_{n-1}(t-s)P(\varphi S(s)f)\,ds \quad (4.19)$$

for every $n \geq 1$, and $S_0(t) = S(t)$ with $\{S(t)\}_{t\geq 0}$ as in (4.15). We now show that the semigroup $\{P(t)\}_{t\geq 0}$ corresponds to the transition function induced by the PDMP ξ as constructed in Sect. 2.3.5. To this end it is enough to show (4.9). Recall from Sect. 2.3.6 that

$$P(t, x, B) = \mathbb{P}_x(\xi(t) \in B) = \lim_{n\to\infty} T_n(t)\mathbf{1}_B(x), \quad x \in X, \ B \in \Sigma, \ t > 0,$$

where $T_n(t)$ is defined in (2.33). Since $T_0(t)\mathbf{1}_B(x) = \mathbf{1}_B(\pi(t, x))\Phi_x(t)$, $x \in X$, Eq. (4.16) implies that

$$\int_X T_0(t)\mathbf{1}_B(x)f(x)\,m(dx) = \int_X \mathbf{1}_B(x)S(t)f(x)\,m(dx)$$

for all $f \in L^1_+$. It follows from (4.19) that

$$\int_0^t \int_X T_0(s)(\varphi T(T_0(t-s)\mathbf{1}_B))(x)f(x)\,m(dx)\,ds$$

$$= \int_0^t \int_X \mathbf{1}_B(x)S_0(t-s)P(\varphi S(s)f)(x)\,m(dx)\,ds = \int_X \mathbf{1}_B(x)S_1(t)f(x)\,m(dx)$$

and, by (2.34), we obtain

$$\int_X T_n(t)\mathbf{1}_B(x)f(x)\,m(dx) = \int_X \mathbf{1}_B(x)\sum_{k=0}^{n} S_k(t)f(x)\,m(dx)$$

for $n = 1$, which is easy to extend to all n, by induction. The dominated convergence theorem and the convergence in (4.19) give

$$\int_X P(t, x, B)f(x)\,m(dx) = \lim_{n\to\infty} \int_X P_n(t, x, B)f(x)\,m(dx)$$

$$= \lim_{n\to\infty} \int_B \sum_{k=0}^{n} S_k(t)f(x)\,m(dx) = \int_B Pf(x)\,m(dx).$$

4.2.4 Dynamical Systems with Jumps

We consider a differential equation

$$x'(t) = b(x(t)) \tag{4.20}$$

on some Borel subset $X \subseteq R^d$ with non-empty interior and with boundary of the Lebesgue measure zero. We assume that b is a locally Lipschitz function on X and that for each $x_0 \in X$ Eq. (4.20) with initial condition $x(0) = x_0$ has a solution $x(t) \in X$ for all $t > 0$ and denote it by $\pi_t(x_0)$. We assume that a point moves according to this flow, i.e. if $x_0 \in X$ is the initial position of a point, then $x = \pi_t(x_0)$ is its location at time t. The point can jump with intensity which is a bounded continuous function $\varphi \colon X \to [0, \infty)$. Let a transition probability function $P(x, B)$ describe the new position of the point after a jump from x and let $L^1 = L^1(X, \mathscr{B}(X), m)$ where m is the Lebesgue measure on X. We assume that $P(x, B)$ satisfies (2.4) and defines a stochastic operator P on L^1, which corresponds to the transition probability $P(x, B)$. If the random variable $\xi(t)$ is the position of the moving point at time t and $\xi(0)$ has density $u(0)$, then $\xi(t)$ has density $u(t)$ which satisfies the evolution equation

$$u'(t) = Au(t) = A_0 u(t) - \varphi u(t) + P(\varphi u(t)), \quad A_0 u(t) = -\sum_{i=1}^{d} \frac{\partial}{\partial x_i}(b_i u).$$
(4.21)

As in Sect. 4.2.1 we define

$$\lambda = \sup\{\varphi(x) \colon x \in X\}, \quad \bar{P} f(x) = \lambda^{-1}\big(P(\varphi f)(x) + (\lambda - \varphi(x))f(x)\big).$$

Then $\bar{P}$ is a stochastic operator and $A = A_0 - \lambda I + \lambda \bar{P}$. The operator A_0 is the generator of a stochastic semigroup $\{P_0(t)\}_{t \geq 0}$ on $L^1(X, \mathscr{B}(X), m)$ given by

$$P_0(t)f(x) = \begin{cases} f(\pi_{-t}x) \det\left[\dfrac{d}{dx}\pi_{-t}x\right], & \text{if } x \in \pi_t(X), \\ 0, & \text{if } x \notin \pi_t(X), \end{cases}$$

where the mapping $x \mapsto \pi_{-t}(x)$ is the inverse of the one to one mapping $\pi_t \colon X \to \pi_t(X)$. The semigroup $\{P(t)\}_{t \geq 0}$ with the generator A can be given by the Dyson–Phillips expansion (3.15) in Corollary 3.2.

4.2.5 Randomly Switched Dynamical Systems

We consider here the simplest case when the number of different dynamical systems is finite. Precisely, we consider a family of systems of differential equations $x' = b^i(x)$, $i \in I = \{1, \ldots, k\}$, defined on some open set $G \subseteq \mathbb{R}^d$. We assume that all functions $b^i \colon G \to \mathbb{R}^d$ are locally Lipschitz functions on G. Let $X \subseteq G$ be a Borel set with non-empty interior and with boundary of Lebesgue measure zero. Let $x_0 \in X$ and $t \geq 0$. We denote by $\pi_t^i(x_0)$ the solution $x(t)$ of the system $x' = b^i(x)$ with the initial condition $x(0) = x_0$ provided that such a solution exists in the time interval $[0, t]$ and $x(s) \in X$ for $s \in [0, t]$. The state of the system is a pair $(x, i) \in \mathbb{X} = X \times I$ or the *dead state* $*$ which is an extra point outside the space $\mathbb{X}$.

If the system is at state (x, i) then we can jump to the state (x, j) with a bounded and continuous intensity $q_{ji}(x)$. Between two consecutive jumps the trajectory starting from the state (x, i) is deterministic and at time t it reaches the state $(\pi_t^i(x), i)$ or the dead state $*$ if $\pi_t^i(x)$ is not defined and then the system spends the rest of the time in the dead state.

Consider the differential equation

$$\frac{d}{dt}x(t) = b^{i(t)}(x(t)), \quad \frac{d}{dt}i(t) = 0. \tag{4.22}$$

Its solutions are solutions of

$$\frac{d}{dt}(x(t), i(t)) = b(x(t), i(t)), \tag{4.23}$$

where the mapping b defined by $b(x, i) = (b^i(x), 0)$ for (x, i) with $x \in G$ is locally Lipschitz continuous. We take the semiflow $\pi : [0, \infty) \times \mathbb{X} \to \mathbb{X} \cup \{*\}$ to be

$$\pi(t, x, i) = (\pi_t^i(x), i), \quad x \in X, \ i \in I,$$

as long as $\pi_t^i(x)$ exists and belongs to X; otherwise we set $\pi(t, x, i) = *$. The jump rate function $\varphi : \mathbb{X} \to [0, \infty)$ is given by

$$\varphi(x, i) = q_i(x) = \sum_{j \neq i} q_{ji}(x),$$

and the survival function is

$$\Phi_{(x,i)}(t) = \exp\left\{-\int_0^t q_i(\pi_s^i(x))\, ds\right\} \quad \text{for } t < t_*(x, i),$$

and $\Phi_{(x,i)}(t) = 0$, otherwise. The jump distribution P is described at the end of Sect. 2.1.6 with the probability of jump from a state (x, i) to the state (x, j) with $j \neq i$ defined by

$$p_{ji}(x) = \begin{cases} 1, & q_i(x) = 0, \ j \neq i, \\ \frac{q_{ji}(x)}{q_i(x)}, & q_i(x) > 0, \ j \neq i, \end{cases}$$

and the corresponding stochastic operator P on L^1 by

$$Pf(x, i) = \sum_{j \neq i} p_{ij}(x) f(x, j).$$

We can define a family of time homogeneous Markov processes $\xi(t)$ on $\mathbb{X} \cup \{*\}$ setting $\xi(t) = (x(t), i(t))$ or $\xi(t) = *$ if we have entered the dead state before time t.

One can also give an analytic definition of the semigroup $\{P(t)\}_{t\geq 0}$ by using a perturbation theorem and an analytic formula for a substochastic semigroup generated by a single semiflow (see e.g. [96]).

4.2.6 Jumps from Boundaries

In this section we show how the perturbation result from Sect. 3.3.4 can be used in an example with only forced jumps. Consider the Lebowitz–Rubinow model from Sect. 1.7. We take $X = [0, 1) \times (0, 1]$ and $\tilde{X} = \mathbb{R}^2$. The solution of (1.4) with initial condition $(x(0), v(0)) = (x, v)$ is of the form

$$\pi(t, x, v) = (x + tv, v), \quad x, v, t \in \mathbb{R}.$$

We have

$$\Gamma = \{1\} \times (0, 1], \quad t_*(x, v) = \frac{1 - x}{v}, \quad x, v \in (0, 1].$$

The survival function is given by $\Phi_{(x,v)}(t) = 1_{[0, t_*(x,v))}(t)$ for $(x, v) \in X$ and the transition probability satisfies

$$P((1, v), \{0\} \times B) = \int_B P(v, dv') = \alpha \delta_v(B) + \beta \int_B k(v', v)\, dv', \quad B \in \mathscr{B}((0, 1]),$$

where k is an integral probability kernel on $(0, 1)$ and α, β are non-negative constants with $\alpha + \beta = 1$. The extended generator $\tilde{L}$ of the process ξ is given by

$$\tilde{L}g(x, v) = v\frac{\partial g(x, v)}{\partial x}, \quad (x, v) \in [0, 1] \times (0, 1],$$

for all bounded functions which are continuously differentiable in the first variable, supplemented with the boundary condition

$$g(1, v) = \int_0^1 g(0, v')P(v, dv'). \tag{4.24}$$

We show that the process $\xi(t) = (x(t), v(t))$ induces a stochastic semigroup $\{P(t)\}_{t\geq 0}$ on the space $L^1 = L^1(X, \Sigma, m)$, where $\Sigma = \mathscr{B}(X)$, and $m(dx \times dv) = dx \times dv$. The infinitesimal generator of this semigroup is of the form

$$Af(x, v) = -v\frac{\partial f}{\partial x}(x, v) \tag{4.25}$$

and the functions f from the domain of A satisfy the boundary condition

$$v'h(0, v') = \alpha v'h(1, v') + \beta \int_0^1 vk(v', v)h(1, v)\, dv. \tag{4.26}$$

We apply Theorem 4.6. Consider the maximal operator

$$Af(x, v) = -v\frac{\partial f}{\partial x}(x, v), \quad f \in \mathscr{D} = \{f \in L^1 : Af \in L^1\},$$

and the space $\mathscr{D}$ with Sobolev norm

$$\|f\|_A = \|f\| + \|Af\|, \quad f \in \mathscr{D}.$$

We have the Green formula

$$\int_X Af(x, v)\, m(dx, dv) = \int_{\Gamma_0} \gamma_0 f(x, v)\, m_0(dx, dv) - \int_{\Gamma_1} \gamma_1 f(x, v)\, m_1(dx, dv)$$

for $f \in \mathscr{D}$, where the boundaries are

$$\Gamma_a = \{a\} \times (0, 1] \quad \text{and} \quad m_a(dx, dy) = v\delta_a(dx)\, dv, \quad a = 0, 1,$$

and the trace operators $\gamma_a \colon f \mapsto f(a, \cdot)$ for $a = 0, 1$, defined on $(\mathscr{D}, \|\cdot\|_A)$ and with values in $L_a^1 = L^1(\Gamma_a, \mathscr{B}(\Gamma_a), m_a)$, are linear and continuous. We define the boundary operator $\Psi \colon \mathscr{D} \to L_0^1$ by $\Psi f = P(\gamma_1 f)$ and the operator $P \colon L_1^1 \to L_0^1$ by

$$Pf(0, v') = \alpha f(1, v') + \beta\frac{1}{v'}\int_0^1 k(v', v)f(1, v)v\, dv.$$

Observe that P preserves the integral on the positive cone so that we have

$$\int_{\Gamma_0} Ph(x, v)\, m_0(dx, dv) = \int_{\Gamma_1} h(x, v)\, m_1(dx, dv), \quad h \in L_1^1, \ h \geq 0.$$

The inverse operator to the boundary operator $\Psi_0 f = \gamma_0 f$, $f \in \mathscr{D}$, when restricted to the nullspace $\mathscr{N}(\lambda I - A)$ of the operator $(\lambda I - A, \mathscr{D})$ is given by

$$\Psi(\lambda)f_0(x, v) = f_0(0, v)e^{-\frac{\lambda x}{v}}, \quad (x, v) \in X, \ f_0 \in L_0^1.$$

We have $P(\gamma_1\Psi(\lambda)) \colon L_0^1 \to L_0^1$ and

$$\|P\gamma_1\Psi(\lambda)f_0\| = \int_0^1 |P\gamma_1\Psi(\lambda)f_0(0, v)|v\, dv \leq e^{-\lambda}\|f_0\|, \quad f_0 \in L_0^1.$$

The operator

$$A_0 f(x, v) = -v\frac{\partial f}{\partial x}(x, v)$$

with domain

$$\mathscr{D}(A_0) = \{f \in \mathscr{D} : \gamma_0 f = 0\}$$

is the generator of the substochastic semigroup $\{S(t)\}_{t \geq 0}$ given by

$$S(t)f(x, v) = f(x - tv, v)\mathbf{1}_{(tv,1]}(x), \quad f \in L^1.$$

For each $t > 0$ the operator $S(t)$ is the restriction to X of the Frobenius–Perron operator $P_0(t)$ for the transformation $\pi_t : \mathbb{R}^2 \to \mathbb{R}^2$ given by $\pi_t(x, v) = (x + tv, v)$ and defined on $L^1(\mathbb{R}^2, \mathscr{B}(\mathbb{R}^2), m)$.

For any $\lambda > 0$ the resolvent of $(A, \mathscr{D}(A))$ at λ with

$$\mathscr{D}(A) = \{f \in \mathscr{D} : \gamma_0 f = P\gamma_1 f\}$$

is a positive operator and for $f \in \mathscr{D}(A)_+$ we have

$$\int_X Af(x, v)\, dx\, dv = \int_{\Gamma_0} \gamma_0 f(x, v)\, m_0(dx, dv) - \int_{\Gamma_1} \gamma_1 f(x, v)\, m_1(dx, dv)$$

$$= \int_{\Gamma_0} P(\gamma_1 f)(x, v)\, m_0(dx, dv) - \int_{\Gamma_1} \gamma_1 f(x, v)\, m_1(dx, dv) = 0.$$

From Theorem 4.6 it follows that A is the generator of a stochastic semigroup.

It remains to show that the semigroup $\{P(t)\}_{t \geq 0}$ corresponds to the transition function induced by the process ξ. It follows from (4.24) that

$$\int_{\Gamma_1} f(x, v)g(x, v)\, m_1(dx, dv) = \int_{\Gamma_0} g(x, v)Pf(x, v)\, m_0(dx, dv').$$

Using the Green formula, we see that

$$\int_X Af(x, v)g(x, y)\, m(dx, dv) = \int_X f(x, v)\widetilde{L}g(x, v)\, m(dx, dv).$$

It is enough to show that a restriction L of the operator $\widetilde{L}$ is the generator of a positive contraction semigroup on a closed subspace of $B(X)$. To this end we make use of Theorem 4.10 and Lemma 3.4. Let

$$\mathscr{X} = \{g \in B(X) : g(\cdot, v) \in C[0, 1], \ v \in (0, 1]\},$$

where $C[0, 1]$ is the space of continuous functions on $[0, 1]$. Consider the operator $\widetilde{L}$ on the maximal domain

$$\widetilde{\mathscr{D}} = \{g \in \mathscr{X} : g \text{ is continuously differentiable in } x \text{ and } \widetilde{L} \in \mathscr{X}\}.$$

We define the boundary operators $\widetilde{\Psi}_0, \widetilde{\Psi} : B(X) \to B(\Gamma)$, where $\Gamma = \Gamma_1$, by

$$\widetilde{\Psi}_0 g(1, v) = g(1, v), \quad \widetilde{\Psi} g(1, v) = \int_0^1 g(0, v') P(v, dv'), \quad v \in (0, 1], \ g \in B(X).$$

The inverse $\widetilde{\Psi}(\lambda)$ of the operator $\widetilde{\Psi}_0$ restricted to the nullspace of the operator $(\lambda I - \widetilde{L}, \widetilde{\mathscr{D}})$ is given by

$$\widetilde{\Psi}(\lambda) g(x, v) = g(1, v) e^{-\frac{\lambda}{v}(1-x)}.$$

It is a positive operator and the operator $I - \widetilde{\Psi} \widetilde{\Psi}(\lambda)$ is invertible for each $\lambda > 0$. The operator L_0 being the restriction of $\widetilde{L}$ to $\mathscr{N}(\widetilde{\Psi}_0)$ is resolvent positive, since

$$R(\lambda, L_0) g(x, v) = e^{\frac{\lambda x}{v}} \int_x^1 e^{-\frac{\lambda r}{v}} f(r, v) \, dr, \quad g \in \mathscr{X}.$$

Consequently, the restriction L of the extended generator $\widetilde{L}$ defined on

$$\mathscr{D}(L) = \{g \in \mathscr{X} : g \text{ satisfies condition (4.24) and } \widetilde{L} g \in \mathscr{X}\}$$

is resolvent positive. Since the boundary condition (4.24) holds for $g = \mathbf{1}_X$ and $L\mathbf{1}_X = 0$, we conclude from Theorem 4.10 that L generates a positive contraction semigroup.

4.2.7 Semigroups for the Stein Model

In this section we show how we can generate a stochastic semigroup for a particular example of a PDMP described in Sect. 1.11 where we considered the Stein model. As in [91] we introduce an additional 0-phase, which starts at the end of the refractory period and lasts until depolarization takes place. Thus we consider three phases $0, 1, 2$. Recall that in phase 1 the depolarization $x(t)$ decays according to the equation $x'(t) = -\alpha x(t)$ with $\alpha > 0$ and phase 2 denotes the refractory phase of duration t_R (see Fig. 4.1). We take $\widetilde{X} = \mathbb{R} \times \{0, 1, 2\}$. The flow $\pi : \mathbb{R} \times \widetilde{X} \to \widetilde{X}$ is given by

$$\pi(t, x, 0) = (x, 0), \quad \pi(t, x, 1) = (xe^{-\alpha t}, 1), \quad \pi(t, x, 2) = (x + t, 2), \quad t, x \in \mathbb{R}.$$

The process is defined on the space $X = \{(0, 0)\} \cup (-\infty, \theta] \times \{1\} \cup [0, t_R) \times \{2\}$ and the active boundary is $\Gamma = \{(t_R, 2)\}$ with $t_*(x, 2) = t_R - x$ for $x \in [0, t_R)$. We take

$$\Phi_{(x,i)}(t) = e^{-(\lambda_E + \lambda_I)t}, \quad i = 0, 1,$$

and $\Phi_{(x,2)}(t) = \mathbf{1}_{[0, t_*(x,2))}(t)$, so that the jump rate function φ is defined by $\varphi(x, i) = \lambda_E + \lambda_I$ for $i = 0, 1$ and $\varphi(x, 2) = 0$. The jump distribution P is given by

$$P((x, i), B) = p_I \mathbf{1}_B(S_I(x, 1)) + p_E \mathbf{1}_B(S_E(x, 1)), \quad (x, i) \in X, \ i = 0, 1,$$

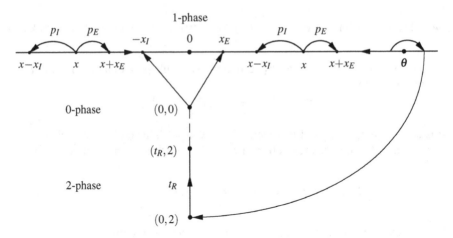

Fig. 4.1 Phases in Stein's model

where

$$p_I = \frac{\lambda_I}{\lambda_E + \lambda_I}, \quad p_E = \frac{\lambda_E}{\lambda_E + \lambda_I}, \quad S_I(x, i) = (x - x_I, 1),$$

and

$$S_E(x, 1) = (x + x_E, 1) \quad \text{for } x \le \theta - x_E, \quad S_E(x, 1) = (0, 2) \quad \text{for } x > \theta - x_E.$$

We take $P((t_R, 2), \{(0, 0)\}) = 1$. The extended generator $\widetilde{L}$ of the process ξ is given by

$$\widetilde{L}g(0, 0) = -(\lambda_E + \lambda_I)g(0, 0) + \lambda_I g(-x_I, 1) + \lambda_E g(x_E, 0),$$

$$\widetilde{L}g(x, 1) = -\alpha x \frac{\partial g(x, 1)}{\partial x} - (\lambda_E + \lambda_I)g(x, 1) + \lambda_I g(x - x_I, 1)$$
$$\qquad + \lambda_E g(x + x_E, 1)\mathbf{1}_{(-\infty, \theta]}(x + x_E) + \lambda_E g(0, 2)\mathbf{1}_{(\theta, \infty)}(x + x_E),$$

$$\widetilde{L}g(x, 2) = \frac{\partial g(x, 2)}{\partial x},$$

for all bounded functions which are differentiable in the first variable, supplemented with the boundary condition

$$g(t_R, 2) = g(0, 0).$$

To find the stochastic semigroup on L^1 we apply Theorem 4.6. Let m be a measure on the σ-algebra $\mathscr{B}(X)$ of the Borel subsets of X such that $m(\{(0, 0)\}) = \delta_{(0,0)}$, the Dirac measure at $(0, 0)$, m restricted to $(-\infty, \theta] \times \{1\}$ is the product of the Lebesgue measure on the interval $(-\infty, \theta]$ and the Dirac measure at $\{1\}$ and m restricted to

$[0, t_R] \times \{2\}$ is the product of the Lebesgue measure on the interval $[0, t_R]$ and the Dirac measure at $\{2\}$. We define the operator P on $L^1 = L^1(X, \mathscr{B}(X), m)$ by

$$Pf(x, 1) = p_E f(x - x_E, 1) + p_I f(x + x_I, 1)\mathbf{1}_{(-\infty, \theta)}(x + x_I),$$

and $Pf(x, 0) = Pf(x, 2) = 0$. We consider the process separately on $(-\infty, -x_I) \times \{1\}$, $[-x_I, x_E] \times \{1\}$ and $(x_E, \theta] \times \{1\}$. We define the maximal operator

$$A_1 f(0, 0) = -(\lambda_E + \lambda_I)f(0, 0),$$

$$A_1 f(x, 1) = \frac{\partial(\alpha x f(x, 1))}{\partial x} - (\lambda_E + \lambda_I)f(x, 1),$$

$$A_1 f(x, 2) = -\frac{\partial f(x, 2)}{\partial x},$$

with domain $\mathscr{D}$ which consists of all functions $f \in L^1$ such that $A_1 f \in L^1$ and f is absolutely continuous on the segments $(-\infty, -x_I) \times \{1\}$, $(-x_I, x_E) \times \{1\}$, $(x_E, \theta) \times \{1\}$, $(0, t_R) \times \{2\}$ and $f(\theta^-, 1) = 0$, where $f(\theta^-, 1)$ is the left limit at θ of the absolutely continuous function $f(x, 1)$. Note that we have

$$\int_{-\infty}^{\theta} A_1 f(x, 1)\, dx = -\alpha x_I f(-x_I^-, 1) + \alpha x_I f(-x_I^+, 1) + \alpha x_E f(x_E^-, 1) - \alpha x_E f(x_E^+, 1)$$

$$- (\lambda_E + \lambda_I) \int_{-\infty}^{\theta} f(x, 1)\, dx$$

and

$$\int_0^{t_R} A_1 f(x, 2)\, dx = f(0^+, 2) - f(t_R^-, 2).$$

We take

$$\Gamma_- = \{(0, 2), (x_E, 1), (-x_I, 1)\}, \quad m_- = \delta_{(0,2)} + \alpha x_E \delta_{(x_E,1)} + \alpha x_I \delta_{(-x_I,1)},$$

and

$$\Gamma_+ = \{(t_R, 2), (x_E, 1), (-x_I, 1)\}, \quad m_+ = \delta_{(t_R,2)} + \alpha x_E \delta_{(x_E,1)} + \alpha x_I \delta_{(-x_I,1)}.$$

Note that Γ_+ contains all points through which the flow π can exit the segments $(-\infty, -x_I) \times \{1\}$, $(-x_I, x_E) \times \{1\}$, $(x_E, \theta) \times \{1\}$, $(0, t_R) \times \{2\}$, in a finite positive time, while the set Γ_- contains points through which the flow can exit these segments in a finite negative time. We define the traces $\gamma_- : \mathscr{D} \to L^1(\Gamma_-)$ by

$$\gamma_- f(0, 2) = f(0^+, 2), \quad \gamma_- f(x_E, 1) = f(x_E^-, 1), \quad \gamma_- f(-x_I, 1) = f(-x_I^+, 1),$$

and $\gamma_+ : \mathscr{D} \to L^1(\Gamma_+)$ by

$$\gamma_+ f(t_R, 2) = f(t_R^-, 2), \quad \gamma_+ f(x_E, 1) = f(x_E^+, 1), \quad \gamma_+ f(-x_I, 1) = f(-x_I^-, 1).$$

The operator

$$A_0 f = A_1 f, \quad f \in \mathscr{D}(A_0) = \{ f \in \mathscr{D} : \gamma_- f = 0 \},$$

is the generator of a substochastic semigroup on L^1. The inverse operator $\Psi(\lambda)$ to the boundary operator $\Psi_0 f = \gamma_- f$ restricted to the nullspace $\mathscr{N}(\lambda I - A_1)$ is given by

$$\Psi(\lambda) f(0,0) = 0, \quad \Psi(\lambda) f(x, 2) = f(0, 2) e^{-\lambda x}, \quad f \in L^1(\Gamma_-),$$

and

$$\Psi(\lambda) f(x, 1) = \begin{cases} f(x_E, 1) \left(\frac{x}{x_E} \right)^{r-1}, & x \in (0, x_E) \\ f(-x_I, 1) \left(\frac{x}{-x_I} \right)^{r-1}, & x \in (-x_I, 0) \\ 0, & x \in (-\infty, -x_I) \cup (x_E, \theta), \end{cases}$$

where $r = (\lambda + \lambda_E + \lambda_I) / \alpha$. The boundary operator Ψ is

$$\Psi f(x, 1) = \gamma_+ f(x, 1) + a(x, 1) f(0, 0), \quad \Psi f(0, 2) = \lambda_E \int_{\theta - x_E} f(x, 1) \, dx,$$

where

$$a(-x_I, 1) = \frac{\lambda_I}{\alpha x_I}, \quad a(x_E, 1) = \frac{\lambda_E}{\alpha x_E}.$$

We have $\Psi\Psi(\lambda) f(x_E, 1) = \Psi\Psi(\lambda) f(-x_I, 1) = 0$ and

$$\Psi\Psi(\lambda) f(0, 2) = \lambda_E \int_{\theta - x_E}^{\theta} \Psi(\lambda) f(x, 1) \, dx$$

for $f \in L^1(\Gamma_-)$. Thus, if $f \geq 0$ then

$$0 \leq \Psi\Psi(\lambda) f(0, 2) \leq \frac{\lambda_E}{\lambda + \lambda_E + \lambda_I} \alpha x_E f(x_E, 1) \leq \frac{\lambda_E}{\lambda + \lambda_E + \lambda_I} \| f \|_{L^1(\Gamma_-)},$$

which implies that $\lambda \| \Psi\Psi(\lambda) \| \leq \lambda_E$. From Theorem 4.6 it follows that the operator $(A_1, \mathscr{N}(\Psi - \Psi_0))$ is the generator of a substochastic semigroup.

The operator

$$Af(0, 0) = A_1 f(0, 0) + f(t_R^-, 2),$$
$$Af(x, i) = A_1 f(x, i) + (\lambda_E + \lambda_I) Pf(x, i), \quad i = 1, 2,$$

with domain

$$\mathscr{D}(A) = \{f \in \mathscr{D} : \Psi_0(f) = \Psi(f)\}$$

is closed. Note that if $f \in \mathscr{D}(A)_+$ then we have

$$\int_X Af(x, i)\, m(dx, di) = 0,$$

since

$$A_1 f(0, 0) + \int_{-\infty}^{\theta} A_1 f(x, 1)\, dx = -(\lambda_E + \lambda_I) \int_{-\infty}^{\theta} f(x, 1)\, dx,$$

and

$$f(t_R^-, 2) + \int_0^{t_R} A_1 f(x, 2)\, dx = \lambda_E \int_{\theta - x_E} f(x, 1)\, dx.$$

It follows from Theorem 4.9 that $(A, \mathscr{D}(A))$ is the generator of a stochastic semi-group.

It remains to show that a restriction of the extended generator $\widetilde{L}$ of the process ξ generates a positive semigroup on a sufficiently large subspace of $B(X)$. We consider the subspace $\mathscr{X} \subset B(X)$ of all functions g such that g restricted to $(-\infty, \theta - x_E) \times \{1\}$ is uniformly continuous function on $(-\infty, \theta - x_E]$, g restricted to $(\theta - x_E, \theta] \times \{1\}$ belongs to $C[\theta - x_E, \theta]$ and $g(\cdot, 2)$ belongs to $C[0, t_R]$. We define $\widetilde{L}_1$ on the domain

$$\mathscr{D}(\widetilde{L}_1) = \{g \in \mathscr{X} : g(t_R, 2) = g(0, 0),\ g \text{ is differentiable in } x,\ \widetilde{L}_1 g \in \mathscr{X}\}$$

by

$$\widetilde{L}_1 g(0, 0) = -(\lambda_E + \lambda_I) g(0, 0),$$

$$\widetilde{L}_1 g(x, 1) = -\alpha x \frac{\partial g(x, 1)}{\partial x} - (\lambda_E + \lambda_I) g(x, 1),$$

$$\widetilde{L}_1 g(x, 2) = \frac{\partial g(x, 2)}{\partial x}.$$

It is easy to see that the operator $\widetilde{L}_1$ is resolvent positive. In particular, we have

$$R(\lambda, \widetilde{L}_1) g(x, 2) = e^{\lambda(x - t_R)} g(0, 0) + e^{\lambda x} \int_x^{t_R} e^{-\lambda y} g(y, 2)\, dy$$

and

$$R(\lambda, \widetilde{L}_1) g(x, 1) = e^{r(\theta - x_E - x)} g((\theta - x_E)^-, 1) + \frac{1}{\alpha} e^{-rx} \int_{\theta - x_E}^x \frac{1}{y} e^{ry} g(y, 1)\, dy$$

for $x > \theta - x_E$, where $r = (\lambda + \lambda_E + \lambda_I)/\alpha$. We can write $\widetilde{L}$ as $\widetilde{L}_1 + \widetilde{B}$, where

$$\widetilde{B}g(x, i) = \lambda_I g(x - x_I, 1) + \lambda_E \left(g(x + x_E, 1)\mathbf{1}_{(-\infty, \theta - x_E]}(x) + g(0, 2)\mathbf{1}_{(\theta - x_E, \theta]}(x) \right)$$

for $(x, i) \in X$, $i = 0, 1$, and $\widetilde{B}g(x, 2) = 0$ for $(x, 2) \in X$. We consider the operator $\widetilde{B}$ on the domain

$$\mathscr{D}(\widetilde{B}) = \{g \in \mathscr{X} : g \text{ is continuous}\}.$$

Since $\widetilde{B}$ is a positive operator and the operator $I - \widetilde{B}R(\lambda, \widetilde{L}_1)$ is invertible for all sufficiently large λ, the operator $L = \widetilde{L}$ with domain $\mathscr{D}(L) = \mathscr{D}(\widetilde{L}_1)$ is resolvent positive. We have $\mathbf{1}_X \in \mathscr{D}(L)$ and $L\mathbf{1}_X = 0$. It follows from Theorem 4.10 that L generates a positive contraction semigroup on the closure of $\mathscr{D}(L)$.

Chapter 5
Asymptotic Properties of Stochastic Semigroups—General Results

In the theory of stochastic processes a special role is played by results concerning the existence of invariant densities and the long-time behaviour of their distributions. These results can be formulated and proved in terms of stochastic semigroups induced by these processes. We consider two properties: asymptotic stability and sweeping. A stochastic semigroup induced by a stochastic process is asymptotically stable if the densities of one-dimensional distributions of this process converge to a unique invariant density. Sweeping is an opposite property to asymptotic stability and it means that the probability that trajectories of the process are in a set Z goes to zero. The main result presented here shows that under some conditions a substochastic semigroup can be decomposed into asymptotically stable parts and a sweeping part. This result and some irreducibility conditions allow us to formulate theorems concerning asymptotic stability, sweeping and the Foguel alternative. This alternative says that under suitable conditions a stochastic semigroup is either asymptotically stable or sweeping.

5.1 Asymptotic Stability and Sweeping

5.1.1 Definitions of Asymptotic Stability and Sweeping

Let the triple (X, Σ, m) be a σ-finite measure space. In the whole chapter, we will consider substochastic operators and semigroups on $L^1 = L^1(X, \Sigma, m)$ corresponding to transition kernels. The *support* of any measurable function f is defined up to a set of measure zero by the formula

$$\text{supp} f = \{x \in X \colon f(x) \neq 0\}.$$

© The Author(s) 2017
R. Rudnicki and M. Tyran-Kamińska, *Piecewise Deterministic Processes in Biological Models*, SpringerBriefs in Mathematical Methods, DOI 10.1007/978-3-319-61295-9_5

Recall that a positive semigroup $\{P(t)\}_{t \geq 0}$ is stochastic if and only if $P^*(t)\mathbf{1}_X = \mathbf{1}_X$ for $t \geq 0$, where $P^*(t)$ is the adjoint of $P(t)$. A density f^* is called *invariant* with respect to a substochastic semigroup if $P(t)f^* = f^*$ for $t \geq 0$. Observe that if a substochastic semigroup $\{P(t)\}_{t \geq 0}$ has an invariant density $f^* > 0$ a.e. then $\{P(t)\}_{t \geq 0}$ is a stochastic semigroup. Indeed,

$$\int_X P^*(t)\mathbf{1}_X(x)f^*(x)\,m(dx) = \int_X \mathbf{1}_X(x)P(t)f^*(x)\,m(dx) = \int_X f^*(x)\,m(dx),$$

but since $P^*(t)\mathbf{1}_X \leq \mathbf{1}_X$ and $f^* > 0$ a.e., we conclude that $P^*(t)\mathbf{1}_X = \mathbf{1}_X$ for all $t \geq 0$.

A substochastic semigroup $\{P(t)\}_{t \geq 0}$ is called *quasi-asymptotically stable* if there exist an invariant density f^* and a positive linear functional α defined on $L^1(X, \Sigma, m)$ such that

$$\lim_{t \to \infty} \| P(t)f - \alpha(f)f^* \| = 0 \quad \text{for} \quad f \in L^1. \tag{5.1}$$

If the semigroup satisfies (5.1) with $\alpha(f) = 1$ for $f \in D$, then it is called *asymptotically stable* [62]. If a substochastic semigroup $\{P(t)\}_{t \geq 0}$ is asymptotically stable then it is a stochastic semigroup, and the notions of quasi-asymptotic stability and asymptotic stability are equivalent for stochastic semigroups.

If the semigroup $\{P(t)\}_{t \geq 0}$ is generated by an evolution equation $u'(t) = Au(t)$ then the asymptotic stability of $\{P(t)\}_{t \geq 0}$ means that the stationary solution $u(t) = f^*$ is asymptotically stable in the sense of Lyapunov and this stability is global on the set D.

A substochastic semigroup $\{P(t)\}_{t \geq 0}$ is called *sweeping* [58] with respect to a set $Z \in \Sigma$ if for every $f \in L^1$ we have

$$\lim_{t \to \infty} \int_Z P(t)f(x)\,m(dx) = 0. \tag{5.2}$$

We usually consider sweeping from a family of sets, e.g. from sets of finite measure m or from all compact subsets of X if X is a topological space.

Sweeping is also called *zero type* or *null property* [79]. Asymptotic stability is also called *strong mixing* or *exactness*, especially in the context of ergodic theory [62]. We can also introduce quasi-asymptotic stability, asymptotic stability and sweeping of a substochastic operator P if in (5.1) and (5.2) we replace $P(t)$ with the iterates P^t of the operator P.

5.1.2   Lower Function Theorem

There are several results concerning asymptotic stability [62]. Some of them are based on assumptions concerning boundedness from below of a stochastic semigroup. One of them is the *lower function theorem* of Lasota and Yorke [63]. A function $h \in L^1$, $h \geq 0$ and $h \neq 0$, is called a *lower function* for a stochastic semigroup $\{P(t)\}_{t \geq 0}$ if

$$\lim_{t \to \infty} \|(P(t)f - h)^-\| = 0 \quad \text{for every } f \in D. \tag{5.3}$$

Here we use the notation $f^-(x) = 0$ if $f(x) \geq 0$ and $f^-(x) = -f(x)$ if $f(x) < 0$. Condition (5.3) can be equivalently written as: there are functions $\varepsilon(t) \in L^1$ such that $\lim_{t \to \infty} \|\varepsilon(t)\| = 0$ and $P(t)f \geq h - \varepsilon(t)$. Observe that if the semigroup is asymptotically stable then its invariant density f^* is a lower function for it. Lasota and Yorke [63] proved the following converse result.

Theorem 5.1 *Let $\{P(t)\}_{t \geq 0}$ be a stochastic semigroup. If there exists a lower function h for a stochastic semigroup $\{P(t)\}_{t \geq 0}$ then this semigroup is asymptotically stable.*

5.1.3 Partially Integral Semigroups and Asymptotic Stability

A substochastic semigroup $\{P(t)\}_{t \geq 0}$ is called *partially integral* (or *with a kernel minorant*) if there exists a measurable function $k \colon (0, \infty) \times X \times X \to [0, \infty)$, called a *kernel*, such that

$$P(t)f(x) \geq \int_X k(t, x, y) f(y) m(dy) \tag{5.4}$$

for every density f and

$$\int_X \int_X k(t, x, y) m(dy) m(dx) > 0$$

for some $t > 0$. If equality holds in (5.4) then $\{P(t)\}_{t \geq 0}$ is said to be an *integral semigroup*. The following theorem is only valid for continuous-time semigroups.

Theorem 5.2 *Let $\{P(t)\}_{t \geq 0}$ be a partially integral stochastic semigroup. Assume that the semigroup $\{P(t)\}_{t \geq 0}$ has a unique invariant density f^*. If $f^* > 0$ a.e., then the semigroup $\{P(t)\}_{t \geq 0}$ is asymptotically stable.*

The proof of Theorem 5.2 is given in [87] and it is based on the theory of Harris operators [38, 54].

We now formulate corollaries which are often used in applications. A substochastic semigroup $\{P(t)\}_{t \geq 0}$ is called *irreducible* if

$$\int_0^\infty P(t)f \, dt > 0 \quad \text{a.e. for every density } f.$$

We say that a substochastic semigroup $\{P(t)\}_{t \geq 0}$ *overlaps supports*, if for every $f_1, f_2 \in D$ there exists $t > 0$ such that

$$m(\text{supp } P(t)f_1 \cap \text{supp } P(t)f_2) > 0.$$

Corollary 5.1 *A partially integral and irreducible stochastic semigroup which has an invariant density is asymptotically stable.*

Corollary 5.2 *A partially integral stochastic semigroup which overlaps supports and has an invariant density $f^* > 0$ a.e. is asymptotically stable.*

Remark 5.1 The assumption in Theorem 5.2 that the invariant density is unique can be replaced by an equivalent one: there does not exist a set $E \in \Sigma$ such that $m(E) > 0$, $m(X \setminus E) > 0$ and $P(t)E = E$ for all $t > 0$. Here $P(t)$ is the operator acting on the σ-algebra Σ defined by: if $f \geq 0$, supp $f = A$ and supp $P(t)f = B$ then $P(t)A = B$. The definition of $P(t)A$ does not depend on the choice of f because if supp $f =$ supp g a.e. then supp $P(t)f =$ supp $P(t)g$ a.e. Irreducibility is equivalent to the following: the semigroup has no nontrivial invariant subset. We recall that a measurable set A is *invariant* with respect to $\{P(t)\}_{t \geq 0}$ if $P(t)(A) \subseteq A$ for $t \geq 0$ (in terms of the transition kernel it means that $P(t, x, X \setminus A) = 0$ for m-a.e. $x \in A$ and $t \geq 0$).

5.1.4　Sweeping via the Existence of σ-finite Invariant Function

We now present results concerning the sweeping property which are based on an assumption about the existence of a σ-finite positive invariant function. These were proved in [58] and in [95] for discrete-time stochastic semigroups (iterates of a stochastic operator) but similar results also hold for continuous-time stochastic semigroups. Before formulating some results we need to extend a substochastic operator beyond the space L^1. If P is a substochastic operator and f is an arbitrary non-negative measurable function, then we define Pf as a pointwise limit of the sequence Pf_n, where (f_n) is any monotonic sequence of non-negative functions from L^1 pointwise convergent to f almost everywhere (see [38, Chap. I]). Observe that we can also define an extension of a substochastic operator using a transition kernel. If $P(x, B)$ is the transition kernel then for any non-negative measurable function f we have $Pf = d\mu/dm$, where $\mu(B) = \int_X f(x)P(x, B)m(dx)$ and $d\mu/dm$ is the Radon–Nikodym derivative.

The following condition plays a crucial role in results concerning sweeping:

(KT) There exists a measurable function f^* such that: $0 < f^* < \infty$ a.e., $P(t)f^* \leq f^*$ for $t \geq 0$, $f^* \notin L^1$ and $\int_Z f^*(x)m(dx) < \infty$ for some set $Z \in \Sigma$ with $m(Z) > 0$.

Theorem 5.3 *([58]) Let $\{P(t)\}_{t \geq 0}$ be an integral substochastic semigroup which has no invariant density. Assume that the semigroup $\{P(t)\}_{t \geq 0}$ and a set $Z \in \Sigma$ satisfy condition (KT). Then the semigroup $\{P(t)\}_{t \geq 0}$ is sweeping with respect to Z.*

Unfortunately, substochastic semigroups induced by PDMPs are usually not integral and we need versions of Theorem 5.3 without assumption (KT). The problem is discussed in details in [95] and we present here only some conclusions. In order to formulate the next theorem we need some auxiliary definitions.

We observe that if a substochastic semigroup $\{P(t)\}_{t \geq 0}$ has a subinvariant function $f^* > 0$ then we can define a new substochastic semigroup $\{\bar{P}(t)\}_{t \geq 0}$ on the space $L^1(X, \Sigma, \mu)$ with $d\mu = f^* dm$ given by

$$\bar{P}(t)f = \frac{1}{f^*} P(t)(ff^*) \tag{5.5}$$

and $\bar{P}(t)1_X \leq 1_X$. Thus, all operators $\bar{P}(t)$ and $\bar{P}^*(t)$ are contractions on $L^1(X, \Sigma, \mu)$ and $L^\infty(X, \Sigma, \mu)$, respectively, and, consequently, they are positive contractions on $L^2(X, \Sigma, \mu)$. Let us define

$$K = \{f \in L^2 : \|\bar{P}(t)f\|_{L^2} = \|\bar{P}^*(t)f\|_{L^2} = \|f\|_{L^2} \ \text{ for } t \geq 0\}$$

and

$$\Sigma_1 = \{B \in \Sigma : 1_B \in K\}.$$

Let $\{P(t)\}_{t \geq 0}$ be a partially integral substochastic semigroup with a kernel $k(t, x, y)$. If

$$\int_X \int_0^\infty k(t, x, y) \, dt \, m(dy) > 0 \quad x - \text{a.e.},$$

then $\{P(t)\}_{t \geq 0}$ is called a *pre-Harris semigroup*.

The following result was proved in [95].

Theorem 5.4 *Let* $\{P(t)\}_{t \geq 0}$ *be a substochastic semigroup which has no invariant density. Assume that the semigroup* $\{P(t)\}_{t \geq 0}$ *and a set* $Z \in \Sigma$ *satisfy condition* (KT). *If one of the following conditions holds:*

(a)   the family Σ_1 *is atomic,*
(b) $\{P(t)\}_{t \geq 0}$ overlaps supports,
(c) $\{P(t)\}_{t \geq 0}$ is a pre-Harris semigroup,

then the semigroup $\{P(t)\}_{t \geq 0}$ *is sweeping with respect to Z.*

The main problem with application of Theorem 5.4 is to find a positive subinvariant function. There are some theoretical results concerning this subject. A substochastic semigroup can have no strictly positive subinvariant function. But under additional assumptions related to the notions of conservative and dissipative semigroups we can find a positive subinvariant function. We precede a formulation of some results with a definition. Let $\{P(t)\}_{t \geq 0}$ be a substochastic semigroup, $f > 0$ be a density, and

$$C = \{x \in X : \int_0^\infty P(t)f(x) \, dt = \infty\}.$$

The definition of C does not depend on the choice of f. The substochastic semigroup $\{P(t)\}_{t \geq 0}$ is called *conservative* if $C = X$ a.e. and *dissipative* if $C = \emptyset$ a.e. From this definition it follows immediately that if $\{P(t)\}_{t\geq 0}$ has an invariant density $f^* > 0$ a.e., then $\{P(t)\}_{t\geq 0}$ is conservative. We also have $P^*(t)\mathbf{1}_C \geq \mathbf{1}_C$ for $t > 0$ or, equivalently, $P(t, x, C) = 1$ for $x \in C$. In particular, a conservative substochastic semigroup is a stochastic semigroup. A substochastic semigroup can be neither conservative nor dissipative, but if it is an irreducible semigroup then it has one of these properties. If a stochastic semigroup is dissipative then for any positive density f the function $f^*(x) = \int_0^\infty P(t)f(x)\,dt$ is a positive and subinvariant function. The function f^* is also not integrable because in the opposite case the semigroup will be conservative. If $\{P(t)\}_{t\geq 0}$ is a conservative pre-Harris semigroup, then $\{P(t)\}_{t\geq 0}$ has a strictly positive invariant function. Even if we know that a semigroup has an subinvariant function f^*, a problem still remains because we need to identify sets Z such that $\int_Z f^*(x)\,m(dx) < \infty$. In order to do this we need to have more information about f^*—not only its existence. We cannot apply directly the formula $Af^* \leq 0$ to find f^* as in the case of a density because $f^* \notin \mathscr{D}(A)$, where A is the infinitesimal generator of the semigroup and $\mathscr{D}(A)$ is its domain.

5.2 Asymptotic Decomposition of Stochastic Semigroups

The main problem with applying results of Sects. 5.1.3 and 5.1.4 to study asymptotic stability is that we need to have an invariant density and a positive subinvariant function to check asymptotic stability and sweeping, respectively. Now we present a general result concerning asymptotic decomposition of substochastic semigroups and some corollaries which do not have any assumption on invariant functions. We hope that these results will prove useful to study asymptotic stability and sweeping of semigroups induced by PDMPs.

5.2.1 Theorem on Asymptotic Decomposition

From now on, we assume that X is a separable metric space and $\Sigma = \mathscr{B}(X)$ is the σ-algebra of Borel subsets of X. We will consider partially integral substochastic semigroups $\{P(t)\}_{t\geq 0}$ which satisfy the following condition:

(K) For every $y_0 \in X$ there exist an $\varepsilon > 0$, a $t > 0$ and a measurable function $\eta \geq 0$ such that $\int \eta(x)\,m(dx) > 0$ and

$$k(t, x, y) \geq \eta(x)\mathbf{1}_{B(y_0,\varepsilon)}(y) \quad \text{for } x \in X, \tag{5.6}$$

where $B(y_0, \varepsilon) = \{y \in X : \rho(y, y_0) < \varepsilon\}$.

In other words condition (K) holds if for each point $x_0 \in X$ we find its neighbourhood U and time $t > 0$ such that for all points $x \in U$ the transition kernel $P(t, x, \cdot)$ can be bounded below by the same nontrivial absolutely continuous measure. Condition (K) was proposed in [95] as a tool to study sweeping of stochastic operators from compact sets. Condition (K) is satisfied if, for example, for every point $y \in X$ there exist a $t > 0$ and an $x \in X$ such that the kernel $k(t, \cdot, \cdot)$ is continuous in a neighbourhood of (x, y) and $k(t, x, y) > 0$.

The following theorem states that a substochastic semigroup can be decomposed into asymptotically stable and sweeping components.

Theorem 5.5 *Let $\{P(t)\}_{t \geq 0}$ be a substochastic semigroup on $L^1(X, \Sigma, m)$, where X is a separable metric space, $\Sigma = \mathscr{B}(X)$, and m is a σ-finite measure. Assume that $\{P(t)\}_{t \geq 0}$ satisfies* (K). *Then there exist an at most countable set J, a family of invariant densities $\{f_j^*\}_{j \in J}$ with disjoint supports $\{A_j\}_{j \in J}$, and a family $\{\alpha_j\}_{j \in J}$ of positive linear functionals defined on $L^1(X, \Sigma, m)$ such that*

(i) *for every $j \in J$ and for every $f \in L^1(X, \Sigma, m)$ we have*

$$\lim_{t \to \infty} \|\mathbf{1}_{A_j} P(t) f - \alpha_j(f) f_j^*\| = 0, \tag{5.7}$$

(ii) *if $Y = X \setminus \bigcup_{j \in J} A_j$, then for every $f \in L^1(X, \Sigma, m)$ and for every compact set F we have*

$$\lim_{t \to \infty} \int_{F \cap Y} P(t) f(x) \, m(dx) = 0. \tag{5.8}$$

Theorem 5.5 was proved in [89] for stochastic semigroups and generalized in [90] for substochastic semigroups. Theorem 5.5 unifies a part of the theory of substochastic semigroups related to asymptotic stability and sweeping and generalizes some earlier results [85, 95].

Since f_j^* is an invariant density with supp $f_j^* = A_j$, the restricted semigroup $\{P_{A_j}(t)\}_{t \geq 0}$ is a stochastic semigroup. Hence, condition (i) implies that $\{P_{A_j}(t)\}_{t \geq 0}$ is asymptotically stable and $\lim_{t \to \infty} \|P(t)f - f_j^*\| = 0$ for $f \in D$ such that supp $f \subseteq A_j$.

Remark 5.2 The sets A_j, $j \in J$, which occur in the formulation of Theorem 5.5, are not only disjoint but also their closures are disjoint. Indeed, let m_j be the Borel measures on X given by

$$m_j(B) = m(B \cap A_j) \quad \text{for } B \in \Sigma.$$

Denote by F_j the *topological support* of the measure m_j, $j \in J$. We recall that the topological support of m_j is the smallest closed set F_j such that $m_j(X \setminus F_j) = 0$ or equivalently $F_j := \{x \in X : m_j(B(x, \varepsilon)) > 0 \text{ for all } \varepsilon > 0\}$. We check that $F_{j_1} \cap F_{j_2} = \emptyset$ for $j_1 \neq j_2$. Suppose that, on the contrary, there is $y_0 \in F_{j_1} \cap F_{j_2}$. Let $t > 0$, $\varepsilon > 0$, and a function η in inequality (5.6) be related to y_0. Fix $j \in \{j_1, j_2\}$. Let f_j^* be the invariant density with the support A_j. From (5.6) it follows that

$$P(t) f_j^*(x) \geq \eta(x) \int_{B(y_0, \varepsilon)} f_j^*(y) \, m(dy) \quad \text{for } x \in X.$$

Since $y_0 \in F_j$, we have $m(B(y_0, \varepsilon) \cap A_j) > 0$ and so $\int_{B(y_0, \varepsilon)} f_j^*(y) \, m(dy) > 0$. This implies that

$$\operatorname{supp} \eta \subseteq \operatorname{supp} P(t) f_j^* = \operatorname{supp} f_j^* = A_j.$$

Hence $\operatorname{supp} \eta \subseteq A_{j_1} \cap A_{j_2} = \emptyset$, a contradiction. Thus $F_{j_1} \cap F_{j_2} = \emptyset$ for $j_1 \neq j_2$. Since F_j is the topological support of m_j, we have $m(A_j \cap F_j) = m(A_j)$ and consequently we can assume that $A_j \subseteq F_j$. Since any metric space is a normal topological space, every two disjoint closed sets of X have disjoint open neighbourhoods. It means that for $j_1, j_2 \in J$ and $j_1 \neq j_2$ we find also disjoint open sets U_{j_1} and U_{j_2} such that $A_{j_1} \subseteq F_{j_1} \subseteq U_{j_1}$ and $A_{j_2} \subseteq F_{j_2} \subseteq U_{j_2}$.

5.2.2 Sweeping and the Foguel Alternative

Theorem 5.5 has a number of corollaries concerning asymptotic stability and sweeping. We start with a corollary concerning sweeping:

Corollary 5.3 *Let $\{P(t)\}_{t \geq 0}$ be a substochastic semigroup on $L^1(X, \Sigma, m)$, where X is a separable metric space, $\Sigma = \mathscr{B}(X)$, and m is a σ-finite measure. Assume that $\{P(t)\}_{t \geq 0}$ satisfies (K) and has no invariant density. Then $\{P(t)\}_{t \geq 0}$ is sweeping from compact sets.*

Condition (K) and irreducibility allow us to express asymptotic properties in the form of the *Foguel alternative*, i.e. the stochastic semigroup is asymptotically stable or sweeping [17]. We use the notion of the Foguel alternative in a narrow sense, when sweeping is from all compact sets.

Corollary 5.4 *Let X be a separable metric space, $\Sigma = \mathscr{B}(X)$ and m be a σ-finite measure. If $\{P(t)\}_{t \geq 0}$ is a substochastic semigroup on the space $L^1(X, \Sigma, m)$, it satisfies (K) and it is irreducible, then $\{P(t)\}_{t \geq 0}$ is asymptotically stable or sweeping from compact sets.*

Proof If $\{P(t)\}_{t \geq 0}$ has an invariant density f^*, then from irreducibility it follows that $\operatorname{supp} f^* = X$. Thus $\{P(t)\}_{t \geq 0}$ is a stochastic semigroup, J is a singleton and condition (i) of Theorem 5.5 holds with $\alpha(f) = 1$ for $f \in D$. If $\{P(t)\}_{t \geq 0}$ has no invariant density, then it is sweeping from compact sets.

In particular, if $\{P(t)\}_{t \geq 0}$ is a substochastic semigroup with a continuous kernel k and $k(t, x, y) > 0$ for some $t > 0$ and for all $x, y \in X$, then the Foguel alternative holds.

An advantage of the formulation of Corollary 5.4 in the form of an alternative is that: in order to show asymptotic stability of a substochastic semigroup we do not need to prove the existence of an invariant density. It is enough to check that the semigroup is non-sweeping. In particular, if the space X is compact and the substochastic semigroup satisfies (K) then it is asymptotically stable. We can also eliminate the sweeping by constructing Hasminskiĭ function (see Sect. 5.2.4).

Since in many applications substochastic semigroups are not irreducible, we are going to formulate another corollary, similar in spirit to Corollary 5.4, but based on a weaker assumption than irreducibility.

We say that a substochastic semigroup $\{P(t)\}_{t\geq 0}$ is *weakly irreducible* if:
(WI) There exists a point $x_0 \in X$ such that for each $\varepsilon > 0$ and for each density f we have

$$\int_{B(x_0,\varepsilon)} P(t)f(x)\, m(dx) > 0 \quad \text{for some } t = t(\varepsilon, f) > 0. \tag{5.9}$$

If a substochastic semigroup $\{P(t)\}_{t\geq 0}$ satisfies conditions (WI) and (K) then the set J defined as in Theorem 5.5 is an empty set or a singleton.

Corollary 5.5 *Let* $\{P(t)\}_{t\geq 0}$ *be a substochastic semigroup on* $L^1(X, \Sigma, m)$, *where* X *is a separable metric space,* $\Sigma = \mathscr{B}(X)$, *and m is a σ-finite measure. Assume that* $\{P(t)\}_{t\geq 0}$ *satisfies* (WI) *and* (K). *Then the semigroup* $\{P(t)\}_{t\geq 0}$ *is sweeping from compact sets or there exist an invariant density* f^* *with the support* A *and a positive linear functional* α *defined on* $L^1(X, \Sigma, m)$ *such that*

(i)   *for every* $f \in L^1(X, \Sigma, m)$ *we have*

$$\lim_{t\to\infty} \|\mathbf{1}_A P(t)f - \alpha(f)f^*\| = 0, \tag{5.10}$$

(ii)   *if* $Y = X \setminus A$, *then for every* $f \in L^1(X, \Sigma, m)$ *and for every compact set* F *we have*

$$\lim_{t\to\infty} \int_{F\cap Y} P(t)f(x)\, m(dx) = 0. \tag{5.11}$$

As a simple corollary we have

Corollary 5.6 *If* X *is a compact space and a substochastic semigroup* $\{P(t)\}_{t\geq 0}$ *satisfies* (WI) *and* (K), *then it is quasi-asymptotically stable or*

$$\lim_{t\to\infty} \int_X P(t)f(x)\, m(dx) = 0. \tag{5.12}$$

In particular, if X *is a compact space,* $\{P(t)\}_{t\geq 0}$ *is a stochastic semigroup which satisfies* (WI) *and* (K), *then it is asymptotically stable.*

We now study asymptotic properties of a substochastic semigroup in the case when X is not compact. We consider the case when X is a union of a countable family of compact sets invariant with respect to $\{P(t)\}_{t\geq 0}$. The following corollary is a simple consequence of the Foguel alternative for semigroups defined on compact spaces but it is very useful in applications.

Corollary 5.7 *Let* $\{P(t)\}_{t\geq 0}$ *be a substochastic semigroup which satisfies all assumptions of Corollary 5.5 and X be a union of an increasing family* (X_n) *of compact sets. We assume that each set* X_n *is invariant with respect to* $\{P(t)\}_{t\geq 0}$ *and that the set* X_1 *includes some neighbourhood of the point* x_0 *from condition* (WI). *Then* $\{P(t)\}_{t\geq 0}$ *is quasi-asymptotically stable or condition* (5.12) *holds. In particular, if the semigroup* $\{P(t)\}_{t\geq 0}$ *is stochastic then it is asymptotically stable.*

5.2.3   Asymptotic Stability

We now formulate a theorem concerning asymptotic stability, which is a consequence of Theorem 5.5. We need an auxiliary condition. We say that a substochastic semigroup $\{P(t)\}_{t\geq 0}$ is *weakly tight* if:

(WT) There exists $\kappa > 0$ such that

$$\sup_{F\in\mathscr{F}} \limsup_{t\to\infty} \int_F P(t)f(x)\,m(dx) \geq \kappa \tag{5.13}$$

for $f \in D_0$, where D_0 is a dense subset of D and $\mathscr{F}$ is the family of all compact subsets of X.

The following theorem is a combination of Theorem 2 [91] and Corollary 5 [90]. The proof is similar to that of Corollary 5 [90], but we give it for the sake of completeness.

Theorem 5.6 *Let* $\{P(t)\}_{t\geq 0}$ *be a substochastic semigroup on* $L^1(X, \Sigma, m)$, *where X is a separable metric space,* $\Sigma = \mathscr{B}(X)$, *and m is a* σ-*finite measure. Assume that* $\{P(t)\}_{t\geq 0}$ *satisfies conditions* (WI), (K), *and* (WT). *Then the semigroup* $\{P(t)\}_{t\geq 0}$ *is quasi-asymptotically stable. If additionally* $\{P(t)\}_{t\geq 0}$ *is a stochastic semigroup then it is asymptotically stable.*

Proof From condition (WT) it follows that the semigroup $\{P(t)\}_{t\geq 0}$ is non-sweeping from all compact sets. Thus, conditions (i) and (ii) of Corollary 5.5 hold. Let A and Y be the sets defined in Corollary 5.5. We claim that

$$\lim_{t\to\infty} \|1_Y P(t)f\| = 0 \quad \text{for } f \in L^1(X, \Sigma, m). \tag{5.14}$$

It is enough to check (5.14) for densities. Then the function $\beta(t) = \int_Y P(t)f(x)\,m(dx)$ is decreasing. Suppose that, contrary to our claim,

$$\lim_{t\to\infty} \int_Y P(t)f_1(x)\,m(dx) = c > 0 \tag{5.15}$$

for some density f_1. We show that there exists a density $\widetilde{f} \in D_0$ such that condition (5.13) does not hold for $\widetilde{f}$. Indeed, let $\varepsilon = c\kappa/4$. Then for sufficiently large t_0 we have

$$\int_Y P(t_0) f_1(x) \, m(dx) < c + \varepsilon.$$

Let $f_2 = \mathbf{1}_Y P(t_0) f_1$. Then

$$\int_A P(t) f_2(x) \, m(dx) < \varepsilon$$

for $t \geq 0$. We define $f_3 = f_2 / \|f_2\|$. Then f_3 is a density and

$$\int_A P(t) f_3(x) \, m(dx) = \frac{1}{\|f_2\|} \int_A P(t) f_2(x) \, m(dx) < \varepsilon/c = \kappa/4 \qquad (5.16)$$

for $t \geq 0$. Condition (ii) of Corollary 5.5 implies that for each compact set F we have

$$\int_{F \cap Y} P(t) f_3(x) \, m(dx) < \kappa/4 \qquad (5.17)$$

for sufficiently large t. Let us take $\widetilde{f} \in D_0$ such that $\|\widetilde{f} - f_3\| \leq \kappa/4$. From inequalities (5.16) and (5.17) it follows that for each compact set F we have

$$\int_F P(t) \widetilde{f}(x) \, m(dx) < \frac{3\kappa}{4} \qquad (5.18)$$

for sufficiently large t. Thus (WT) does not hold for $\widetilde{f}$, and, consequently, condition (5.14) is fulfilled. Conditions (5.10) and (5.14) imply that the semigroup $\{P(t)\}_{t \geq 0}$ is quasi-asymptotically stable. If the semigroup $\{P(t)\}_{t \geq 0}$ is stochastic, then quasi-asymptotic stability is equivalent to asymptotic stability.

5.2.4 Hasminskiĭ Function

If we know that a stochastic semigroup satisfies the Foguel alternative then we can eliminate sweeping by constructing a Hasminskiĭ function [86]. Now we show how a slight modification of this method can be applied to study weak tightness. It is clear that a weakly tight stochastic semigroup is non-sweeping from compact sets. We precede the definition of a Hasminskiĭ function by some observations.

Let A be a generator of a substochastic semigroup $\{P(t)\}_{t \geq 0}$ and let

$$R := R(1, A) = (I - A)^{-1} = \int_0^\infty e^{-t} P(t) \, dt \qquad (5.19)$$

be the resolvent of the operator A. It is clear that R is a substochastic operator and if $\{P(t)\}_{t \geq 0}$ is a stochastic semigroup, then R is also a stochastic operator.

Lemma 5.1 *Let $\{P(t)\}_{t\geq0}$ be a substochastic semigroup, $f \in D$ and $Z \in \Sigma$. If $\limsup_{n\to\infty} \int_Z R^n f(x)\, m(dx) \geq \varepsilon$ then $\limsup_{t\to\infty} \int_Z P(t) f(x)\, m(dx) \geq \varepsilon$.*

Proof Let $f \in D$. From (5.19) it follows that

$$R^{n+1} f = \int_0^\infty r_n(t) P(t) f \, dt,$$

where $r_n(t) = \dfrac{t^n e^{-t}}{n!}$. Suppose that, on the contrary, there exist $t_0 > 0$ and $\varepsilon' < \varepsilon$ such that

$$\int_Z P(t) f(x)\, m(dx) \leq \varepsilon' \quad \text{for } t \geq t_0.$$

Since $\int_0^\infty r_n(t)\, dt = 1$, $\int_0^{t_0} r_n(t)\, dt \to 0$ as $n \to \infty$, and

$$\int_Z R^n f(x)\, m(dx) \leq \int_0^{t_0} r_n(t)\, dt + \int_{t_0}^\infty r_n(t) \int_Z P(t) f(x)\, m(dx)\, dt,$$

we obtain $\limsup_{n\to\infty} \int_Z R^n f(x)\, m(dx) \leq \varepsilon'$, a contradiction.

Let P be a stochastic operator, $Z \in \Sigma$ and $V: X \to [0, \infty)$ be a measurable function. By D_V we denote the set

$$D_V = \{f \in D: \int_X f(x) V(x)\, m(dx) < \infty\}.$$

Since m is a σ-finite measure, the set D_V is dense in D. Assume that there exist $M > 0$ and $\varepsilon > 0$ such that

$$\int_X V(x) P f(x)\, m(dx) \leq -\varepsilon + \int_X V(x) f(x)\, m(dx) + M \int_Z P f(x)\, m(dx) \quad (5.20)$$

for $f \in D_V$. Then the function V will be called a *Hasminskiĭ function* for the operator P and the set Z.

Lemma 5.2 *Let P be a stochastic operator and $Z \in \Sigma$. Assume that there exists a Hasminskiĭ function for the operator P and the set Z. Then for $f \in D$ we have*

$$\limsup_{n\to\infty} \int_Z P^n f(x)\, m(dx) \geq \varepsilon/2M, \qquad (5.21)$$

where ε and M are the constants from formula (5.20).

Proof Let $f \in D_V$. Then from (5.20) it follows that $P^n f \in D_V$ for every integer $n \geq 1$. Thus

$$\int_X V(x) P^{n+1} f(x)\, m(dx) \le \int_X V(x) P^n f(x)\, m(dx) - \varepsilon + M \int_Z P^{n+1} f(x)\, m(dx)$$

$$(5.22)$$

for $f \in D_V$ and $n \ge 1$. Suppose, contrary to our claim, that (5.21) does not hold. Then there exists a positive integer $n_0 = n_0(f)$ such that $\int_Z P^{n+1} f(x)\, m(dx) < \varepsilon/(2M)$ for $n \ge n_0$. From (5.22) it follows that

$$\int_X V(x) P^{n+1} f(x)\, m(dx) \le \int_X V(x) P^n f(x)\, m(dx) - \frac{\varepsilon}{2} \qquad (5.23)$$

for $n \ge n_0$, and consequently

$$\lim_{n \to \infty} \int_X V(x) P^n f(x)\, m(dx) = -\infty,$$

which is impossible. Thus (5.21) holds for $f \in D_V$. Since the set D_V is dense in D, (5.21) is satisfied for all $f \in D$.

Consider a stochastic semigroup $\{P(t)\}_{t \ge 0}$ with the generator A and let $R = (I - A)^{-1}$. A measurable function $V : X \to [0, \infty)$ is called a *Hasminskiĭ function* for the semigroup $\{P(t)\}_{t \ge 0}$ and a set $Z \in \Sigma$ if (5.20) holds for $P = R$. Lemmas 5.1 and 5.2 imply the following

Corollary 5.8 *Let $\{P(t)\}$ be a stochastic semigroup generated by the equation*

$$u'(t) = Au(t).$$

Assume that there exists a Hasminskiĭ function for this semigroup and a set Z. Then

$$\limsup_{t \to \infty} \int_Z P(t) f(x)\, m(dx) \ge \varepsilon/2M \quad \text{for } f \in D. \qquad (5.24)$$

In particular, the semigroup $\{P(t)\}$ is non-sweeping with respect to the set Z. If Z is a compact set then $\{P(t)\}$ is weakly tight.

Let $P = R$. Using the formula $R = I + AR$ we can write (5.20) in the following form

$$\int_X V(x) AR f(x)\, m(dx) \le -\varepsilon + (\overline{M} + \varepsilon) \int_Z R f(x)\, m(dx) \qquad (5.25)$$

for $f \in D_V$, where $\overline{M} = M - \varepsilon$. Now we assume that there exists a measurable function formally denoted by A^*V such that

$$\int_X A^*V(x) g(x)\, m(dx) = \int_X V(x) Ag(x)\, m(dx) \quad \text{for } g \in \mathscr{D}(A) \cap D_V, \qquad (5.26)$$

$$A^*V(x) \leq \overline{M} \quad \text{for } x \in Z \quad \text{and} \quad A^*V(x) \leq -\varepsilon \quad \text{for } x \in X \setminus Z. \qquad (5.27)$$

Then condition (5.24) holds and the stochastic semigroup $\{P(t)\}_{t\geq 0}$ is non-sweeping with respect to the set Z. If Z is a compact set then $\{P(t)\}_{t\geq 0}$ is weakly tight.

The function V is called the Hasminskiĭ function because he has showed [47] that the semigroup generated by a non-degenerate Fokker–Planck equation has an invariant density if and only if there exists a positive function V such that $A^*V(x) \leq -c < 0$ for $\|x\| \geq r$. The main difficulty in using this method is to define in a proper way A^*V because usually V does not belong to the domain of the operator A^*. The method of Hasminskiĭ function has been applied to continuous-time Markov chains [92], multi-state diffusion processes [86] and diffusion with jumps [88], where inequality (5.20) was proved by using some generalization of the maximum principle. This method was also applied in [84] to flows with jumps (see the end of Sect. 1.5) but the proof of inequality (5.20) is based on an approximation of V by a sequence of elements from the domain of the operator A^*.

Chapter 6
Asymptotic Properties of Stochastic Semigroups—Applications

Most of time-homogeneous piecewise deterministic Markov processes induce stochastic semigroups. Therefore, general analytic tools used in the description of long-time behaviour of stochastic semigroups from Chap. 5 can be useful to study asymptotic and ergodic properties of PDMPs. In this chapter, we give applications of general results concerning stochastic semigroups to PDMPs considered in Chap. 1.

6.1 Continuous-Time Markov Chains

6.1.1 Foguel Alternative

We now study stochastic semigroups on the space $l^1 = L^1(\mathbb{N}, 2^{\mathbb{N}}, m)$, where m is the counting measure, and the elements of l^1 are real-valued sequences $u = (u_i)_{i \in \mathbb{N}}$ such that $\sum_{i=0}^{\infty} |u_i| < \infty$. Thus, the density u is a sequence (u_i) such that $u_i \geq 0$ for $i \in \mathbb{N}$ and $\sum_{i=0}^{\infty} u_i = 1$.

Let $Q = [q_{ij}]$ be a sub-Kolmogorov matrix, as introduced in Sect. 4.2.1. Consider a minimal substochastic semigroup $\{P(t)\}_{t \geq 0}$ induced by the equation

$$u'(t) = Qu(t), \tag{6.1}$$

i.e. the semigroup generated by an extension of the operator Q. Since $X = \mathbb{N}$ is a discrete space, the semigroup $\{P(t)\}_{t \geq 0}$ is integral with a continuous kernel k. If $u_i(0) > 0$ then $u_i(t) > 0$ for sufficiently small $t > 0$ because $\{P(t)\}_{t \geq 0}$ is a C_0-semigroup. Thus $k(t, i, i) > 0$, which means that the kernel k satisfies condition (K) and all assumptions of Theorem 5.5 are fulfilled.

© The Author(s) 2017
R. Rudnicki and M. Tyran-Kamińska, *Piecewise Deterministic Processes in Biological Models*, SpringerBriefs in Mathematical Methods,
DOI 10.1007/978-3-319-61295-9_6

We consider the following irreducibility condition:

(I) for all $i, j \in \mathbb{N}$ there exists a sequence of non-negative integers $i_0, i_1, \ldots, i_r$ such that $i_0 = j$, $i_r = i$ and

$$q_{i_r i_{r-1}} \cdots q_{i_2 i_1} q_{i_1 i_0} > 0. \tag{6.2}$$

Condition (I) implies that the semigroup $\{P(t)\}_{t \geq 0}$ is irreducible and Corollary 5.4 gives the following theorem.

Theorem 6.1 *Let $\{P(t)\}_{t \geq 0}$ be a substochastic semigroup on l^1 induced by (6.1). If (I) holds, then the semigroup satisfies the Foguel alternative:*
(a) if $\{P(t)\}_{t \geq 0}$ has an invariant density, then it is asymptotically stable,
(b) if $\{P(t)\}_{t \geq 0}$ has no invariant density, then for every $u \in l^1$ and $i \in \mathbb{N}$ we have

$$\lim_{t \to \infty} (P(t)u)_i = 0. \tag{6.3}$$

6.1.2 Non-explosive Markov Chains and Asymptotic Stability

We now present a theorem on asymptotic stability for stochastic semigroups induced by Markov chains. If the minimal semigroup $\{P(t)\}_{t \geq 0}$ related to a Kolmogorov matrix Q is a stochastic semigroup, then the matrix Q, the Markov chain and the minimal semigroup are called *non-explosive*.

Theorem 6.2 *Let Q be a non-explosive Kolmogorov matrix. We assume that there exist a sequence $v = (v_i)$ of non-negative numbers and positive constants ε, k, M such that*

$$\sum_{i=0}^{\infty} q_{ij} v_i \leq \begin{cases} M & \text{for } j \leq k, \\ -\varepsilon & \text{for } j > k. \end{cases} \tag{6.4}$$

Then the stochastic semigroup $\{P(t)\}_{t \geq 0}$ related to Q is non-sweeping from the set $\{0, 1, \ldots, k\}$. In particular, if the matrix Q satisfies conditions (I) and (6.4), then the semigroup $\{P(t)\}_{t \geq 0}$ is asymptotically stable.

The proof of Theorem 6.2 based on the idea of Hasminskiĭ function is given in [92, Theorem 10].

As an example of application we consider a birth–death process (see Sect. 1.2). The evolution of densities of this process is given by the following system of equations:

$$u_i'(t) = -a_i u_i(t) + b_{i-1} u_{i-1}(t) + d_{i+1} u_{i+1}(t) \tag{6.5}$$

for $i \geq 0$, where $b_{-1} = d_0 = 0$, $b_i \geq 0$, $d_{i+1} \geq 0$, $a_i = b_i + d_i$ for $i \geq 0$. The matrix Q corresponding to Eq. (6.5) is a Kolmogorov matrix.

Assume that $d_i \geq b_i$ for $i \geq k$, where k is a given positive integer. We check using Theorem 4.11 that the matrix Q is non-explosive. We can choose $\lambda > 0$ such that

$d_i + \lambda \geq b_i$ for $i \in \mathbb{N}$. Consider a sequence $w = (w_i)_{i \geq 0}$ of non-negative numbers satisfying $Q^* w = \lambda w$. Then

$$d_i w_{i-1} - (b_i + d_i) w_i + b_i w_{i+1} = \lambda w_i \qquad (6.6)$$

for all $i \geq 1$, where $w_{-1} := 0$. Then from (6.6) we obtain

$$w_{i+1} - w_i = \left(\frac{d_i + \lambda}{b_i} \right) w_i - \frac{d_i}{b_i} w_{i-1} \geq w_i - w_{i-1}$$

for all $i \in \mathbb{N}$. If w is a non-zero sequence then from the last inequality it follows that $\lim_{i \to \infty} w_i = \infty$. It means that $w \notin l^\infty$ and according to Theorem 4.11 the matrix Q is non-explosive and, consequently, it generates a stochastic semigroup.

We now consider again a birth–death process with $b_i > 0$ and $d_{i+1} > 0$ for all $i \geq 0$. Let us assume that there exists $\varepsilon > 0$ such that $b_i \leq d_i - \varepsilon$ for $i \geq k$. Then the system (6.5) generates a stochastic semigroup and condition (I) holds. Let $v_i = i$ for $i \geq 0$, then

$$\sum_{i=0}^{\infty} v_i q_{ij} = (j-1) d_j - j (b_j + d_j) + (j+1) b_j = b_j - d_j \leq -\varepsilon$$

for $j \geq k$, which implies condition (6.4). Hence, the stochastic semigroup generated by the system (6.5) is asymptotically stable.

6.1.3 Markov Chain with an Absorbing State

We now consider a non-explosive Markov chain such that condition (I) holds for all i, j except of $j = 0$. It means that states $i, j \geq 1$ communicate, $q_{i0} = 0$ for all $i \geq 0$, and for some $j \geq 1$ we have $q_{0j} > 0$. Then 0 is an absorbing state, $u^0 = (1, 0, 0, \dots)$ is an invariant density, and condition (WI) holds for u^0. According to Corollary 5.5 there exists $w \in l^\infty$ with $w_i \geq 0$ for all i such that

(i) for each $u \in l^1$ we have $\lim_{t \to \infty} \| (P(t)u)_0 - \alpha(u) \| = 0$, where $\alpha(u) = \sum_{i=0}^{\infty} u_i w_i$,

(ii) $\lim_{t \to \infty} (P(t)u)_i = 0$ for $i \geq 0$.

Generally, it is not easy to find $\alpha(u)$ but in some cases such formulae are known. As an example, we consider a birth–death process with $b_i > 0$, $d_i > 0$ for $i \geq 1$ and $b_0 = 0$. The assumption $b_0 = 0$ is highly natural because it means that the zero-size population cannot develop. Let $\rho_i = \prod_{j=1}^{i} \frac{d_j}{b_j}$ for $i \geq 1$. Then $w_0 = 1$,

$$w_k = \frac{\sum_{i=k}^{\infty} \rho_i}{1 + \sum_{i=1}^{\infty} \rho_i} \quad \text{if} \quad \sum_{i=1}^{\infty} \rho_i < \infty,$$

and $w_k = 1$ in the opposite case. For example if $b_k = bk^a$, $d_k = dk^a$, where $b > d$, $a \in [0, 1]$, then $\rho_i = (d/b)^i$ and we find that $w_k = (d/b)^k$ for $k = 0, 1, 2 \ldots$.

6.1.4 Asymptotics of Paralog Families

In Sect. 1.18 we considered a PDMP which describes the evolution of paralog families in a genome. We recall that if $s_k(t)$ is the number of k-element families at time t then

$$s_1'(t) = -(d + r)s_1(t) + 2(2m + r)s_2(t) + m \sum_{j=3}^{\infty} j s_j(t), \tag{6.7}$$

$$s_k'(t) = d(k - 1)s_{k-1}(t) - (d + r + m)k s_k(t) + (r + m)(k + 1)s_{k+1}(t) \tag{6.8}$$

for $k \geq 2$. The following result [98] characterizes the long-time behaviour of the sequences $(s_k(t))$.

Theorem 6.3 *Let $\mathscr{X}$ be the space of sequences (s_k) which satisfy the condition $\sum_{k=1}^{\infty} k|s_k| < \infty$. There exists a sequence $(s_k^*) \in \mathscr{X}$ such that for every solution $(s_k(t))$ of (1.35) and (1.36) with $(s_k(0)) \in \mathscr{X}$ we have*

$$\lim_{t \to \infty} e^{(r-d)t} s_k(t) = C s_k^* \tag{6.9}$$

for every $k = 1, 2, \ldots$ and C dependent only on the sequence $(s_k(0))$. Moreover if $d = r$ then

$$\lim_{t \to \infty} s_k(t) = C \frac{\beta^i}{k}, \tag{6.10}$$

where $\beta = \dfrac{r}{r + m}$.

The idea of the proof of Theorem 6.3 is to replace system (6.7)–(6.8) by a new one which generates a stochastic semigroup. Let

$$u_i(t) = e^{(r-d)t} i s_i(t).$$

Then

$$u_1' = -(2d + m)u_1 + (m + r)u_2 + \sum_{k=1}^{\infty} m u_k, \tag{6.11}$$

$$u_k' = -(d + r + m + \tfrac{d-r}{k})k u_k + d k u_{k-1} + (r + m)k u_{k+1} \quad \text{for } i \geq 2. \tag{6.12}$$

This system is of the form $u'(t) = Qu(t)$. It is easy to check that Q is a Kolmogorov matrix. We apply Theorem 4.11 to prove that Q is non-explosive. The case $d = 0$ is trivial, so we can assume that $d > 0$. The sequence $w = (w_k)_{k \geq 1}$ satisfies equation

$Q^*w = \lambda w$ if and only if

$$w_{n+1} = \left(1 + \frac{\lambda}{(n+1)d}\right)w_n + \frac{(n-1)(r+m)}{(n+1)d}(w_n - w_{n-1}) + \frac{m}{(n+1)d}(w_n - w_1)$$

for $n \geq 1$. Hence, the sequence (w_n) is increasing. Thus

$$w_{n+1} \geq \left(1 + \frac{\lambda}{(n+1)d}\right)w_n,$$

and consequently

$$w_n \geq w_1 \prod_{i=2}^{n}\left(1 + \frac{\lambda}{di}\right) \quad \text{for } n \geq 2.$$

Since the product $\prod_{k=1}^{\infty}(1 + \lambda d^{-1}k^{-1})$ diverges, we have $w \notin l^{\infty}$ and according to Theorem 4.11 the matrix Q generates a stochastic semigroup.

We next apply Theorem 5.1 to prove that the semigroup $\{P(t)\}_{t \geq 0}$ generated by the system (6.11)–(6.12) is asymptotically stable. From Eq. (6.11) applied to densities we obtain

$$u_1'(t) \geq -(2d+m)u_1(t) + m.$$

This implies that

$$\liminf_{t \to \infty} u_1(t) \geq \frac{m}{2d+m}. \tag{6.13}$$

Then $h = (\frac{m}{2d+m}, 0, 0, \dots)$ is a lower function and the semigroup is asymptotically stable. Let $u^* = (u_k^*)$ be an invariant density. Then (6.9) holds with $s_k^* = u_k^*/k$. If $r = d$, then the invariant density is of the form $u_k^* = \frac{m}{r}\left(\frac{r}{r+m}\right)^k$, which gives (6.10).

Another proof of Theorem 6.3 can be carried out with the help of Theorem 6.1 by checking that this semigroup is non-sweeping from the set $\{1\}$. A generalization of this result to the case when the mutation rate depends on the family size was made in [99].

6.1.5 Applications to Pure Jump-Type Markov Processes

We now study asymptotic properties of the semigroup $\{P(t)\}_{t \geq 0}$ related to a pure jump-type Markov process $\xi(t)$ as described in Sect. 4.2.1 with a bounded measurable jump rate function φ and the jump distribution P corresponding to a stochastic operator P. We will apply Theorem 5.6, and therefore we require that X is a separable metric space and $\Sigma = \mathscr{B}(X)$. Let us assume that $\inf\{\varphi(x)\colon x \in X\} > 0$. Let $Q(x, B) = \sum_{n=0}^{\infty} P^n(x, B)$, where the sequence of transition probablities $P^n(x, B)$ is given by the induction formula:

$$P^0(x, B) = \delta_x(B), \quad P^{n+1}(x, B) = \int_X P(y, B) P^n(x, dy).$$

We assume that there exists a point $x_0 \in X$ such that for each neighbourhood U of x_0 we have $Q(x, U) > 0$ for x-a.e. Then from formula (4.13) it follows that the semigroup $\{P(t)\}_{t \geq 0}$ is weakly irreducible, i.e. it satisfies condition (WI). Moreover, we suppose that for every $y_0 \in X$ there exist an $\varepsilon > 0$, an $n \in \mathbb{N}$, and a measurable function $\eta \geq 0$ such that $\int \eta(x) m(dx) > 0$ and

$$P^n(y, B) \geq \int_B \eta(x) m(dx) \quad \text{for } y \in B(y_0, \varepsilon) \text{ and } B \in \Sigma. \qquad (6.14)$$

Again from (4.13) the semigroup $\{P(t)\}_{t \geq 0}$ satisfies condition (K). In order to give a condition which guarantees asymptotic stability, we can use a Hasminskiĭ function. Let $V \colon X \to [0, \infty)$ be a measurable function. We define

$$A^*V(x) = -\varphi(x)V(x) + \varphi(x) \int_X V(y) P(x, dy).$$

This definition is formally correct (though it can happen that $A^*V(x) = \infty$ on a set of positive measure m), and formula (5.26) holds. If for some compact set F we have $A^*V(x) \leq M$ for $x \in F$ and $A^*V(x) \leq -\varepsilon$ for $x \in X \setminus F$, then there exists a Hasminskiĭ function for the semigroup $\{P(t)\}_{t \geq 0}$ and the set F, and consequently $\{P(t)\}_{t \geq 0}$ is weakly tight. Since the semigroup $\{P(t)\}_{t \geq 0}$ satisfies conditions (WI), (K) and (WT) of Theorem 5.6, it is asymptotically stable.

Example 6.1 Consider a pure jump-type Markov process on $X = [0, \infty)$ such that we jump from a point x to the point ϑ_x, where the random variable ϑ_x has values in X. We assume that there exist $r > 0$ and $\delta > 0$ such that

$$\sup_{x \leq r} \mathbb{E}\vartheta_x < \infty \text{ for } x < r \quad \text{and} \quad \mathbb{E}\vartheta_x \leq x - \delta \text{ for } x \geq r. \qquad (6.15)$$

We also assume that for each $x \geq 0$ the random variable ϑ_x has a continuous density f_x and that there exists a point $x_0 \in X$ such that $f_x(x_0) > 0$ for $x \leq r$. Since the densities f_x are continuous, condition (K) is fulfilled. From (6.15) we deduce that for each $x \geq r$ there is a point $y < x - \varepsilon$ such that $y \in \text{supp } f_x$. Using induction argument we check that $P^n(x, [0, r]) > 0$ for sufficiently large n, and since $f_x(x_0) > 0$ for $x \leq r$ we finally have $P^{n+1}(x, U) > 0$ for every neighbourhood of x_0, i.e. (WI) holds. We check that $V(x) = x$ is a Hasminskiĭ function. We have

$$A^*V(x) = -\varphi(x)V(x) + \varphi(x) \int_X V(y) P(x, dy) = \varphi(x)(\mathbb{E}\vartheta_x - x).$$

From (6.15) we see that A^*V is bounded from above and $A^*V(x) \leq -\varepsilon$ for $x \geq r$, where $\varepsilon = \delta \inf_{x \in X} \varphi(x)$. Since the semigroup $\{P(t)\}_{t \geq 0}$ satisfies conditions (WI), (K) and (WT), it is asymptotically stable.

6.2 Dynamical Systems with Random Switching

6.2.1 General Formulation

Based on the paper [90] we present here a general method of studying asymptotic properties of semigroups induced by dynamical systems with random switching introduced in Sect. 1.9 and described in Sect. 4.2.5. The family of processes $\xi(t)$ induces a substochastic semigroup $\{P(t)\}_{t \geq 0}$ on the space $L^1(\mathbb{X}, \Sigma, m)$, where Σ is the σ-algebra of the Borel subsets of $\mathbb{X} = X \times I$ and m is the product measure of the Lebesgue measure on X and the counting measure on I. The semigroup $\{P(t)\}_{t \geq 0}$ corresponds to the transition function P given by

$$P(t, (x, i), B) = \text{Prob}(\xi_{x,i}(t) \in B),$$

where $\xi_{x,i}(t)$ is a Markov process from the family $\xi(t)$ starting from the state (x, i).

We now recall how to check condition (K) for the semigroup $\{P(t)\}_{t \geq 0}$. Let $n \in \mathbb{N}$, $\mathbf{i} = (i_1, \ldots, i_n) \in I^n$, $i_p \neq i_{p-1}$ for $p = 2, \ldots, n$, and $\mathbf{t} = (t_1, \ldots, t_n)$ be such that $t_p > 0$ for $p = 1, \ldots, n$. Take $y \in X$ and assume that the sequence $y_0, \ldots, y_n$ given by the recurrent formula $y_0 = y$, $y_p = \pi_{t_p}^{i_p}(y_{p-1})$ for $p = 1, \ldots, n$, is well defined and $q_{i_{p+1} i_p}(y_p) > 0$ for $p = 1, \ldots, n - 1$. The function

$$\pi_{\mathbf{t}}^{\mathbf{i}}(y) = \pi_{t_n}^{i_n} \circ \cdots \circ \pi_{t_1}^{i_1}(y)$$

is called a *cumulative flow* along the trajectories of the flows $\pi^{i_1}, \ldots, \pi^{i_n}$ which joins the state (y, i_1) with the state (y_n, i_n). For $\mathbf{i} \in I^{d+1}$, $y \in X$, and $t > 0$ we define the function $\psi_{y,t}$ on the set $\Delta_t = \{\tau = (\tau_1, \ldots, \tau_d): \tau_i > 0, \tau_1 + \cdots + \tau_d \leq t\}$ by

$$\psi_{y,t}(\tau_1, \ldots, \tau_d) = \pi_{(\tau_1, \ldots, \tau_d, t - \tau_1 - \tau_2 - \cdots - \tau_d)}^{\mathbf{i}}(y).$$

Assume that for some $y_0 \in X$ and $\tau^0 \in \Delta_t$ we have

$$\det \left[\frac{d\psi_{y_0,t}(\tau^0)}{d\tau} \right] \neq 0. \tag{6.16}$$

Then there exists a continuous and positive function κ defined in a neighbourhood V of the point (x_0, y_0), such that $k(t, (x, i_{d+1}), (y, i_1)) \geq \kappa(x, y)$ for $(x, y) \in V$, where

$k(t, \cdot, \cdot)$ is the kernel of $P(t)$ and $x_0 = \psi_{y,t}(\tau_1^0, \ldots, \tau_d^0)$. It means that condition (K) holds for the state (y_0, i_1).

Condition (6.16) can be checked by using Lie brackets. Let $f(x)$ and $g(x)$ be two vector fields on $\mathbb{R}^d$. The *Lie bracket* $[f, g]$ is a vector field given by

$$[f, g]_j(x) = \sum_{k=1}^{d} \left(f_k(x) \frac{\partial g_j}{\partial x_k}(x) - g_k(x) \frac{\partial f_j}{\partial x_k}(x) \right).$$

Assume that the vector fields $b_1, \ldots, b_k$ are sufficiently smooth in a neighbourhood of a point x. We say that *Hörmander's condition* holds at x if vectors fields

$$b^2(x) - b^1(x), \ldots, b^k(x) - b^1(x), \ [b^i, b^j](x)_{1 \le i, j \le k}, \ [b^i, [b^j, b^l]](x)_{1 \le i, j, l \le k}, \ldots$$

span the space $\mathbb{R}^d$. Let $(x_0, i), (y_0, j) \in \mathbb{X}$ and assume that the cumulative flow joins (y_0, j) with (x_0, i). If Hörmander's condition holds at x_0 and $q_{l'l}(x_0) > 0$ for $l \ne l'$, then condition (K) is satisfied at the state (y_0, j) (see [8, Theorem 4]).

We now show how to check condition (WI). Let (x_0, i) be given. We assume that for each $\varepsilon > 0$ and each state (y_0, j) we can find $x \in B(x_0, \varepsilon/2)$ such that (y_0, j) can be joined with (x, i) by a cumulative flow $\pi_{\mathbf{t}^0}^{\mathbf{i}}(y_0)$, where $i_1 = j$, $i_n = i$ and $\mathbf{t}^0 = (t_1^0, \ldots, t_n^0)$. From continuous dependence of solutions on initial conditions and continuity of intensity functions q_{i_{p+1}, i_p} it follows that there exists $\delta > 0$ such that each state (y, j) with $y \in B(y_0, \delta)$ can be joined with a state (x', i) with $x' \in B(x_0, \varepsilon/2)$, i.e. $\pi_{\mathbf{t}^0}^{\mathbf{i}}(y) \in B(x_0, \varepsilon/2)$ for $y \in B(y_0, \delta)$. We now can find $\gamma > 0$ such that $\pi_{\mathbf{t}}^{\mathbf{i}}(y) \in B(x_0, \varepsilon)$ if $\|\mathbf{t} - \mathbf{t}^0\| < \gamma$. It means that

$$\mathrm{Prob}\left(\pi_{\mathbf{t}}^{\mathbf{i}}(y) \in B(x_0, \varepsilon) \text{ and } t_1 + \cdots + t_n = t \right) > 0$$

for $y \in B(y_0, \delta)$ and $t = t_1^0 + \cdots + t_n^0$. Therefore

$$P(t, (y, j), B(x_0, \varepsilon) \times \{i\}) > 0 \quad \text{for } y \in B(y_0, \delta). \tag{6.17}$$

Inequality (5.9) in condition (WI) can be written in the following way:

$$\int_{\mathbb{X}} f(y, j) P(t, (y, j), B(x_0, \varepsilon) \times \{i\}) \, m(dy \times dj) > 0. \tag{6.18}$$

Let (y_0, j) be a point such that

$$m\left((B(y_0, \delta) \times \{j\}) \cap \mathrm{supp} \, f \right) > 0 \quad \text{for each } \delta > 0. \tag{6.19}$$

Then inequality (6.18) follows from (6.17) and (6.19). Thus, condition (WI) holds if there exists a state (x_0, i) such that starting from any state we are able to go arbitrarily close to (x_0, i) by a cumulative flow. For example, if x_0 is an asymptotically stable stationary point of a semiflow π_t^i with the basin of attraction $U \subseteq X$ and each state

(y, j) can be joined by a cumulative flow with some state (x, i), where $x \in U$, then condition (WI) holds.

We now explain how to apply Corollary 5.7 to dynamical systems with random switching. We assume that we have an increasing sequence of bounded open sets $G_n \subseteq G$ with smooth boundaries ∂G_n, such that $X \subseteq \bigcup_{n=1}^{\infty} \mathrm{cl} G_n$, where $\mathrm{cl} G_n$ denotes the closure of G_n. Let $\mathbf{n}(x)$ be an outward pointing normal field of the boundary ∂G_n. We assume that

$$\mathbf{n}(x) \cdot b^i(x) < 0 \quad \text{for } x \in \partial G_n, i \in I, \text{ and } n \in \mathbb{N}. \tag{6.20}$$

Then the solutions of each system $x' = b^i(x)$ starting from points in $\mathrm{cl} G_n$ remain in the set $\mathrm{cl} G_n$. In particular the sets $X_n = X \cap \mathrm{cl} G_n$ are invariant with respect to the semigroup $\{P(t)\}_{t \geq 0}$.

We summarize the above discussion. Consider a family of systems of differential equations $x' = b^i(x)$, $i \in I$, such that for each point $\bar{x} \in X$ the initial problem $x' = b^i(x)$, $x(0) = \bar{x}$, has a unique solution $x(t) \in X$ for all $t \geq 0$. Then a dynamical system with random switching created by this family of systems and some continuous and bounded intensity functions $q_{ji}(x)$ generates a stochastic semigroup $\{P(t)\}_{t \geq 0}$ on $L^1(\mathbb{X}, \mathscr{B}(\mathbb{X}), m)$. Assume that there exist $x_0 \in X$ and $i_0 \in I$ such that starting from any state $(x, i) \in \mathbb{X}$ we are able to go arbitrarily close to (x_0, i_0) by a cumulative flow and that Hörmander's condition holds at x_0 and $q_{l'l}(x_0) > 0$ for all $l, l', l \neq l'$. We also assume that there is an increasing family of bounded open sets $G_n \subseteq \mathbb{R}^d$ such that $X \subseteq \bigcup_{n=1}^{\infty} \mathrm{cl} G_n$ and condition (6.20) holds. Then the semigroup $\{P(t)\}_{t \geq 0}$ is asymptotically stable.

6.2.2 Applications to Stochastic Gene Expression Models

We now examine asymptotic properties of the stochastic models of gene expression (see Sects. 1.8 and 1.9). In order to find sufficient conditions for asymptotic stability of the semigroups related to these models we can apply the procedure described in the last section. It is easy to see that if we substitute $x(t) = \frac{\mu}{P}\xi\left(\frac{t}{\mu}\right)$ and $i(t) = A\left(\frac{t}{\mu}\right)$ to (1.14) then we obtain

$$x'(t) = i(t) - x(t). \tag{6.21}$$

Similarly, substituting $x_1(t) = \frac{\mu_R}{R}\xi_1\left(\frac{t}{\mu_R}\right)$, $x_2(t) = \frac{\mu_P \mu_R}{PR}\xi_2\left(\frac{t}{\mu_R}\right)$, $i(t) = A\left(\frac{t}{\mu_R}\right)$, $\alpha = \frac{\mu_P}{\mu_R}$ to (1.15) gives

$$\begin{cases} x_1'(t) = i(t) - x_1(t), \\ x_2'(t) = \alpha(x_1(t) - x_2(t)). \end{cases} \tag{6.22}$$

After a simple linear substitution we can also replace (1.16) by

$$\begin{cases} x_1'(t) = i(t) - x_1(t), \\ x_2'(t) = \alpha(x_2(t) - x_1(t)), \\ x_3'(t) = \beta(x_3(t) - x_2(t)). \end{cases} \tag{6.23}$$

From now on we consider only models given by (6.21), (6.22) and (6.23). In all these models we have two dynamical systems, which correspond, respectively, to $i = 0$ and $i = 1$. Each system has a unique stationary point, either $\mathbf{0} = (0, 0, 0)$, or $\mathbf{1} = (1, 1, 1)$, which is asymptotically stable. Assume that intensities q_0 and q_1 of transformations of a gene into an active state and into inactive state, respectively, are non-negative continuous functions of x. We assume that $q_0(\mathbf{0}) > 0$ and $q_1(\mathbf{1}) > 0$. This assumption implies two properties of the relative PDMP. The first one is that if the process starts from any point $x \in \mathbb{R}^d$ then it enters the invariant set $X = [0, 1]^d \times \{0, 1\}$ and the process visits any neighbourhood of the point $(\mathbf{0}, 0)$ and the point $(\mathbf{1}, 1)$ for infinitely many times. It means that condition (WI) holds for $x^0 = \mathbf{0}$ and for $x^0 = \mathbf{1}$. Moreover, if we assume additionally that $q_0(\mathbf{1}) > 0$ or $q_1(\mathbf{0}) > 0$ then we can check that condition (K) holds by applying Hörmander's condition. We consider here only the most advanced three-dimensional model. Let $v = (1, 0, 0)$. Then $b^2(x) = v + b^1(x)$. Hence

$$b^2 - b^1 = (1, 0, 0),$$
$$[b^1, b^2] = [b^1, v] = [(-x_1, \alpha(x_2 - x_1)), \beta(x_3 - x_2)), (1, 0, 0)] = (1, -\alpha, 0),$$
$$[b^1, [b^1, b^2]] = (1, -(\alpha^2 + \alpha), \alpha\beta).$$

The vectors $b^2 - b^1$, $[b^1, b^2]$, $[b^1, [b^1, b^2]]$ span the space $\mathbb{R}^3$, and consequently Hörmander's condition holds. Since the set X is compact, the semigroup $\{P(t)\}_{t\geq 0}$ is asymptotically stable according to Corollary 5.6.

The one-dimensional model from Sect. 1.8 is simple enough to find the invariant density, which allows us to give more information on asymptotic behaviour of distributions of this process. In this case the functions $u_i(t, x) = P(t)f(x, i)$, $i \in \{0, 1\}$, satisfy the following Fokker–Planck system:

$$\frac{\partial u_0}{\partial t} + \frac{\partial}{\partial x}(-xu_0) = q_1 u_1 - q_0 u_0,$$
$$\frac{\partial u_1}{\partial t} + \frac{\partial}{\partial x}((1 - x)u_1) = q_0 u_0 - q_1 u_1.$$

If u_i, $i \in \{0, 1\}$, do not depend on t, then

$$\begin{aligned} (-xu_0(x))' &= q_1 u_1 - q_0 u_0, \\ ((1 - x)u_1(x))' &= q_0 u_0 - q_1 u_1. \end{aligned} \tag{6.24}$$

Hence

$$-xu_0(x) + (1 - x)u_1(x) = c \quad \text{for some } c \in \mathbb{R}.$$

From the last equation we find $u_1(x) = \dfrac{c}{1-x} + \dfrac{x}{1-x} u_0(x)$. If we suppose that $u(x, i)$ is a density then both functions u_0 and u_1 are non-negative and integrable, but this case is possible only if $c = 0$. Indeed, if $c < 0$ then

$$\int_0^x \frac{s u_0(s)}{1-s}\, ds \le x \int_0^x \frac{u_0(s)}{1-s}\, ds < \frac{|c| x}{2}$$

for sufficiently small $x > 0$, and consequently u_1 cannot be a non-negative function. If $c > 0$, then u_1 is not integrable in 1. Substituting $u_1(x) = \dfrac{x}{1-x} u_0(x)$ to (6.24) we get

$$u_0'(x) = a(x) u_0(x), \quad a(x) = \frac{q_0(x) - 1}{x} - \frac{q_1(x)}{1-x}.$$

Thus

$$u_0(x) = C \exp \int_0^x \left[\frac{q_0(s) - 1}{s} - \frac{q_1(s)}{1-s} \right] ds,$$

$$u_1(x) = \frac{Cx}{1-x} \exp \int_0^x \left[\frac{q_0(s) - 1}{s} - \frac{q_1(s)}{1-s} \right] ds,$$

(6.25)

where we choose C such that $\int_0^1 (u_0(x) + u_1(x))\, dx = 1$.

Example 6.2 Consider the case when q_0 and q_1 are constants. Then

$$u_0(x) = C x^{q_0 - 1} (1-x)^{q_1} \quad \text{and} \quad u_1(x) = C x^{q_0} (1-x)^{q_1 - 1}.$$

This pair is a density if

$$C = (B(q_0 + 1, q_1) + B(q_0, q_1 + 1))^{-1},$$

where B is the Beta function.

Remark 6.1 Observe that if $q_0(0) > 0$ and $q_1(1) > 0$ then the functions u_0 and u_1 given by (6.25) are integrable. That means that there exists a unique density for this semigroup and since the semigroup is partially integral, it is asymptotically stable according to Theorem 5.2. Also in two- and three-dimensional cases the semigroup remains asymptotically stable when $q_0(1) = 0$ and $q_1(0) = 0$, but the proof is more advanced because we need to check directly condition (6.16).

Remark 6.2 It is rather difficult to find a general formula for an invariant density in two- and three-dimensional cases. But we can find a formula for the support of invariant density if q_0 and q_1 are strictly positive functions (see [100]). Also in this paper we can find a discussion concerning *adiabatic limit* of the system, i.e. the behaviour of our model, when both rates q_0 and q_1 tend to infinity, unlike their ratio. In this case, our process "tends" to a deterministic model, which can exhibit two specific types of behaviour: bistability and the existence of the limit cycle.

Remark 6.3 If $q_0(0) = 0$ or $q_1(1) = 0$ then the system can loose asymptotic stability even in the case when $q_i(x)$ are positive for all other x. Consider one-dimensional example with $q_0(x) = x^r$, $r > 0$ and $q_1(x) = 1$. Then from (6.25) it follows that $u_0(x) = C\frac{1-x}{x} \exp\left(r^{-1}x^r\right)$ and $u_1(x) = C \exp\left(r^{-1}x^r\right)$. In this case u_0 is not integrable in 0, and therefore an invariant density does not exist. In this example the distributions of the process converge weakly to $\delta_{(0,0)}$. It means that the average time when the gene is inactive tends to 1. If $q_0(0) = 0$ and $q_1(1) = 0$, then the distribution of the process can converge to $c_1\delta_{(0,0)} + c_2\delta_{(1,1)}$, where the constants c_1 and c_2 can depend on the initial density. Such properties can be checked by using Theorem 5.3.

6.3 Cell Maturation Models

6.3.1 Introductory Remarks

In this section we consider size-structured (or maturity-structured) cellular mathematical models. Examples of such models were presented in Sects. 1.5, 1.6, 1.7 and 1.13. Generally, two types of such models are considered. The first group consists of discrete-time models, which show us how the size or maturation is inherited from a mother to a daughter cell. Such models are popular in the description of the cell cycle. The second type includes continuous-time models, when we observe the evolution of size distribution. We should underline that there are a lot of papers and books concerning such models [76]. Some of them are very advanced. For example, a continuous-time two-phase model [68] is given by a system of nonlinear partial differential equations with time and space retarded arguments. We present only the simplest models which can be easily treated using PDMPs.

The models presented in Sects. 1.5 and 1.6 belong to the category of dynamical systems with random jumps, i.e. we have one dynamical system and a point moves along the trajectories of this system and with some intensity jumps to a new place and resumes movement. In a single-phase model of cell cycle, we have a jump at the moment of cellular division, but in two or more phase models we can have also jumps when a cell enter the next phase. Such models are based on the assumption that in each generation and each phase the size or maturation changes according to the same law. Moreover, since we consider a single line of cells we do not take into account birth and death rates which usually depends on cell parameters and the size of population. At the end of this section we give some remarks how methods developed for stochastic semigroups can be applied to positive semigroups related to more general size-structured models.

6.3.2 Flows with Jumps

We now generalize the method considered for pure jump-type Markov processes to flows with jumps described in Sect. 4.2.4. We show how we can apply Theorem 5.6 to prove asymptotic stability of the stochastic semigroup $\{P(t)\}_{t\geq 0}$. Define $R(t, x, B) = \varphi(\pi_t x) P(\pi_t x, B)$ for $t \geq 0$, $x \in X$, $B \in \Sigma$ and by induction

$$R(t_1, \ldots, t_n, t_{n+1}, x, B) = \int_X R(t_{n+1}, y, B) R(t_1, \ldots, t_n, x, dy).$$

Let $Q(t_1, t_2, \ldots, x, B) = \sum_{n=0}^{\infty} R(t_1, \ldots, t_n, x, B)$ for any sequence of non-negative numbers (t_n). We fix a point $x_0 \in X$ and assume that for a.e. point $x \in X$ and each neighbourhood U of x_0 we find a sequence (t_n) such that $Q(t_1, t_2, \ldots, x, U) > 0$. Then from formulae (3.15) and (3.14) it follows that the semigroup $\{P(t)\}_{t\geq 0}$ satisfies condition (WI).

We now discuss the problem how to check condition (K). If for example the transition function P has a continuous kernel minorant $k \geq 0$, i.e. $P(x, B) \geq \int_B k(x, y) \, dy$, and for every point $x \in X$ we find $t \geq 0$ and $y \in X$ such that $k(\pi_t x, y) > 0$, then (K) holds.

But in many applications the transition function can have no kernel part. Consider an example when the jump is deterministic from x to $S(x)$, i.e. $P(x, \{S(x)\}) = 1$. We assume that the transformation S is continuously differentiable and the Frobenius–Perron operator P_S exists (see Sect. 2.1.5). Let $x_0 \in X$ and define a sequence $x_j = S(x_{j-1})$ for $j = 1, \ldots, d$. Set

$$v_j = S'(x_{d-1}) \ldots S'(x_{j-1}) b(x_{j-1}) - b(x_d) \tag{6.26}$$

for $j = 1, \ldots, d$. We say that the point x_0 satisfies Hörmander's condition if $\varphi(x_{j-1}) > 0$ for all $j = 1, \ldots, d$ and the vectors $v_1, \ldots, v_d$ are linearly independent. We say that the point $y_0 \in X$ *can be joined with* $\tilde{x} \in X$ if there is a sequence $t_0, t_1, \ldots, t_n$ of positive numbers such that $y_{i+1} := S(\pi_{t_i} y_i)$, $\varphi(\pi_{t_i} y_i) > 0$ for $i = 0, \ldots, n-1$, and $\tilde{x} = \pi_{t_n} y_n$. Assume that each point $y_0 \in X$ can be joined with a point $\tilde{x}$ which satisfies Hörmander's condition. Then condition (K) holds (see [87] the proof of Proposition 1). Moreover, if we find one point x_0 such that each point y_0 can be joined with x_0 then condition (WI) holds. If for any $x, y \in X$ we can join x with y, then the semigroup $\{P(t)\}_{t\geq 0}$ is irreducible.

In order to give a condition which guarantees asymptotic stability, we can use a Hasminskiĭ function. Let $V : X \to [0, \infty)$ be a continuously differentiable function. We define

$$A^*V(x) = A_0^*V(x) - \varphi(x)V(x) + \varphi(x) \int_X V(y) \, P(x, dy),$$

where $A_0^* V(x) = \sum_{i=1}^{d} b_i(x) \dfrac{\partial V}{\partial x_i}(x)$. If for some compact set F we have $A^* V(x) \leq M$
for $x \in F$ and $A^* V(x) \leq -\varepsilon$ for $x \in X \setminus F$, then V is a Hasminskiĭ function for
the semigroup $\{P(t)\}_{t\geq 0}$ and the set F. The proof of this fact can be found in [84]
and it is based on an approximation of V by a sequence of elements from the domain
of the operator A^*. Thus $\{P(t)\}_{t\geq 0}$ is weakly tight. Since the semigroup $\{P(t)\}_{t\geq 0}$
satisfies conditions (WI), (K) and (WT) of Theorem 5.6, it is asymptotically stable.

6.3.3   Size-Structured Model

We now investigate a size-structured model from Sect. 1.5. We keep the notation from
Sect. 1.5, i.e. x is the size of a cell, it changes according to the equation $x' = g(x)$;
$\varphi(x)$ is the intensity of division; if the mother cell has size x at the moment of division,
then the new born daughter cell has size $x/2$. Here the jump is deterministic and given
by $S(x) = x/2$. Let $x_{\min}$ be the minimum cell size. We assume that φ is a continuous
function and that $\varphi(x) = 0$ for $x \leq 2x_{\min}$ and $\varphi(x) > 0$ for $x > 2x_{\min}$, because $2x_{\min}$
is the minimum cell size when it can divide. Thus we can consider our process in the
phase space $X = [x_{\min}, \infty)$. We also assume that the function $g \colon X \to (0, \infty)$ has
a bounded derivative. Moreover, we assume that

$$g(2\bar{x}) \neq 2g(\bar{x}) \quad \text{for some } \bar{x} > x_{\min}. \tag{6.27}$$

Then taking $x_0 = 2\bar{x}$ and $x_1 = S(x_0) = \bar{x}$ condition (6.26) takes the form

$$v_1 = S'(x_0)g(x_0) - g(x_1) = \tfrac{1}{2}g(2\bar{x}) - g(\bar{x}) \neq 0.$$

We can join any point $y_0 \in X$ with x_0 by the action of the dynamical system and
transformation S. Indeed, if $y_0 < x_0$ then there exists $t > 0$ such that $x_0 = \pi_t(y_0)$.
If $y_0 > x_0$ then $S^n(y_0) < x_0$ for some positive integer n and, therefore, it can be
joined with x_0. Hence condition (K) holds. The semigroup $\{P(t)\}_{t\geq 0}$ induced by this
process is irreducible because we can join any two points from the set $(x_{\min}, \infty)$.

From Corollary 5.4 it follows that the semigroup $\{P(t)\}_{t\geq 0}$ is asymptotically stable
or sweeping from compact sets. Therefore, if $\{P(t)\}_{t\geq 0}$ has an invariant density f^*,
then $\{P(t)\}_{t\geq 0}$ is asymptotically stable and $f^* > 0$ a.e.; if $\{P(t)\}_{t\geq 0}$ has no invariant
density, then for all $c > x_{\min}$ and $f \in L^1$ we have

$$\lim_{t\to\infty} \int_{x_{\min}}^{c} P(t)f(x)\,dx = 0.$$

The evolution equation is of the form

$$\frac{\partial u}{\partial t} = -\frac{\partial}{\partial x}(g(x)u(t, x)) - \varphi(x)u(t, x) + 2\varphi(2x)u(t, 2x). \tag{6.28}$$

Thus, if there exists an invariant density f^*, then it satisfies the following equation:

$$g(x)f^{*\prime}(x) = -(\varphi(x) + g'(x))f^*(x) + 2\varphi(2x)f^*(2x). \tag{6.29}$$

Since (6.29) is an ordinary differential equations with advanced argument, it is rather difficult to find its solution or to check when a density is one of its solutions (see [70] for sufficient conditions and particular examples).

We now apply the Hasminskiĭ function technique to find a sufficient condition for asymptotic stability. Assume that

$$\limsup_{x\to\infty} \varphi(x) > 0, \quad \text{and} \quad \kappa = \limsup_{x\to\infty} \frac{g(x)}{x\varphi(x)} < \log 2. \tag{6.30}$$

Then we check that there exists a Hasminskiĭ function. We have

$$A^*V(x) = g(x)V'(x) - \varphi(x)V(x) + \varphi(x)V(\tfrac{1}{2}x).$$

Let $V(x) = x^r$, where $r > 0$. Then

$$A^*V(x) = x^r[rx^{-1}g(x) - (1 - 2^{-r})\varphi(x)]. \tag{6.31}$$

Let $h(r) = (1 - 2^{-r})/r$. Since $\lim_{r\to 0^+} h(r) = \log 2$, we find $r > 0$ such that $h(r) > \kappa$. From (6.30) and (6.31) it follows that there exists $\varepsilon > 0$ such that $A^*V(x) \le -\varepsilon$ for sufficiently large x. Thus if (6.27) and (6.30) hold, then the semigroup $\{P(t)\}_{t\ge 0}$ is asymptotically stable.

Remark 6.4 Age- and size-structured models play an important role in population dynamics. The first one was introduced in [74, 105] and the second in [12]. Both models were intensively studied and generalized [31, 46, 76, 97, 118]. In [11] it is considered an evolution equation $u' = Au$ which can be used to study both types of models. This equation generates a semigroup of positive operators $\{T(t)\}_{t\ge 0}$ on some space $L^1(X, \Sigma, \mu)$. The main result of this paper is *asynchronous exponential growth* of the population. Precisely, under suitable assumptions there exist $\lambda \in \mathbb{R}$, an integrable function f^* and a positive linear functional α such that

$$\lim_{t\to\infty} e^{-\lambda t}T(t)f = \alpha(f)f^*. \tag{6.32}$$

We can use the technique developed for stochastic semigroup in Chap. 5 to prove (6.32). In order to do it we first prove that there exist $\lambda \in \mathbb{R}$ and function w such that $A^*w = \lambda w$ and $0 < c_1 \le w(x) \le c_2$ for $x \in X$. Then we introduce a stochastic semigroup $\{P(t)\}_{t\ge 0}$ on the space $L^1(X, \Sigma, m)$ with $m(dx) = w(x)\mu(dx)$ given by $P(t) = e^{-\lambda t}T(t)$. We check that $\{P(t)\}_{t\ge 0}$ is asymptotically stable, which gives (6.32).

6.3.4 Two-Phase Cell Cycle Model

We now consider the two-phase cell cycle model presented in Sect. 1.6. This is a simplified version of the model introduced in [68]. We assume that the size x of a cell changes according to the equation $x' = g(x)$, where the function g has a bounded derivative; the intensity of division $\varphi(x)$ is a continuous function, $\varphi(x) = 0$ for $x \le x_B$ and $\varphi(x) > 0$ for $x > x_B$, where x_B is the minimum cell size when it can enter the phase B; the constant $t_B > 0$ is the duration of the phase B. The minimum cell size is $x_{\min} = \frac{1}{2}\pi_{t_B}x_B$ and it is clear that we need to assume that $\pi_{t_B}x_B < 2x_B$ to have $x_{\min} < x_B$. Thus we can consider our process on the space $X = [x_{\min}, \infty) \times \{1\} \cup [x_B, \infty) \times [0, t_B] \times \{2\}$. As in Sect. 6.3.3 we assume that

$$g(2\bar{x}) \ne 2g(\bar{x}) \quad \text{for some } \bar{x} > x_{\min}. \tag{6.33}$$

Since this model is not exactly a special case of the general model considered in Sect. 6.3.2, we explain how to check condition (K). Let $\bar{z} = \pi_{-t_B}(2\bar{x})$. Then $\bar{z} > x_B$ and $\varphi(\bar{z}) > 0$. First, we check that (K) holds for $y_0 = (\bar{z}, 1)$. Precisely, we show that there exist $t > t_B, \varepsilon > 0, x_0 \in (x_{\min}, \infty) \times \{1\}$ and neighbourhoods U of y_0 and V of x_0, such that the operator $P(t)$ is partially integral with the kernel

$$k(x, y) \ge \varepsilon \mathbf{1}_V(x)\mathbf{1}_U(y). \tag{6.34}$$

Let $z > x_B$ and $t > t_B$. We introduce a function $\psi_{z,t}$ defined on the interval $[t_B, t]$ by

$$\psi_{z,t}(\tau) = \pi_{t-\tau}\left(\tfrac{1}{2}\pi_\tau z\right).$$

We have

$$\frac{d\psi_{z,t}}{d\tau}(\tau) = -g\left(\pi_{t-\tau}\left(\tfrac{1}{2}\pi_\tau z\right)\right) + \frac{\partial \pi_{t-\tau}}{\partial x}\left(\tfrac{1}{2}\pi_\tau z\right) \cdot \tfrac{1}{2}g(\pi_\tau z),$$

$$\lim_{t \to t_B^+} \lim_{z \to \bar{z}} \frac{d\psi_{z,t}}{d\tau}(t_B) = -g\left(\tfrac{1}{2}\pi_{t_B}\bar{z}\right) + \tfrac{1}{2}g\left(\pi_{t_B}\bar{z}\right) = -g(\bar{x}) + \tfrac{1}{2}g(2\bar{x}) \ne 0.$$

Therefore we find $\delta > 0$ and $M > 0$ such that

$$\delta < \left|\frac{d\psi_{z,t}}{d\tau}(\tau)\right| < M \tag{6.35}$$

for z from some neighbourhood U_1 of $\bar{z}$, for some $t > t_B$ and $\tau \in [t_B, t]$. Now we consider our stochastic process starting from the point $(z, 1)$. Let the random variable T be the first time when the process jumps to the phase B. Since $\varphi(\bar{z}) > 0$, T has a density distribution h such that $h(\tau) \ge \beta$ for $\tau \in [0, t - t_B]$, where β is a positive constant. Let $\tau_0 = \frac{1}{2}(t_B + t)$, $x_0 = \left(\pi_{t-\tau_0}\left(\tfrac{1}{2}\pi_{\tau_0}\bar{z}\right), 1\right)$ and $V = (x_0 - r, x_0 + r) \times \{1\}$ with $r = \frac{1}{2}\delta(t - t_B)$. Then (6.34) holds with $\varepsilon = \beta/M$. Since every point $x \in X$ can be joined with $(\bar{z}, 1)$, condition (K) is fulfilled. The semigroup $\{P(t)\}_{t \ge 0}$ induced by this process also satisfies condition (WI) because we can join any point from X with y_0. The infinitesimal generator of $\{P(t)\}_{t \geq 0}$ is of the following form:

$$
\begin{aligned}
Af(x, 1) &= -\frac{\partial}{\partial x}(g(x)f(x, 1)) - \varphi(x)f(x, 1) + 2f(2x, t_B, 2), \\
Af(x, \tau, 2) &= -\frac{\partial}{\partial x}(g(x)f(x, \tau, 2)) - \frac{\partial}{\partial \tau}(f(x, \tau, 2)),
\end{aligned}
\tag{6.36}
$$

and the functions from the domain of A satisfy the condition

$$
\varphi(x)f(x, 1) = f(x, 0, 2).
$$

We find that

$$
\begin{aligned}
A^*f(x, 1) &= g(x)\frac{\partial f}{\partial x}(x, 1) - \varphi(x)f(x, 1) + \varphi(x)f(x, 0, 2), \\
A^*f(x, \tau, 2) &= g(x)\frac{\partial f}{\partial x}(x, \tau, 2) + \frac{\partial f}{\partial \tau}(x, \tau, 2),
\end{aligned}
\tag{6.37}
$$

and the functions from the domain of A^* satisfy

$$
f(x, 1) = f(2x, t_B, 2). \tag{6.38}
$$

We take $V(x, 1) = x^r$ and $V(x, \tau, 2) = Ce^{-a\tau}x^r$, where r, a and C are some positive constants. Then

$$
\begin{aligned}
A^*V(x, 1) &= x^r[rg(x)x^{-1} + \varphi(x)(C - 1)], \\
A^*V(x, \tau, 2) &= Ce^{-a\tau}x^r[rg(x)x^{-1} - a].
\end{aligned}
$$

From (6.38) we find that $C = 2^{-r}e^{at_B}$. We assume that

$$
\limsup_{x \to \infty} \frac{g(x)}{x} < \kappa < \frac{\log 2}{t_B}, \quad \limsup_{x \to \infty} \frac{g(x)}{x\varphi(x)} < \log 2 - \kappa t_B.
$$

Then taking $a = \kappa r$, we check that there exist $\varepsilon > 0$ and a sufficiently small $r > 0$ such that $A^*V(x, 1) \leq -\varepsilon$ and $A^*V(x, \tau, 1) \leq -\varepsilon$ for $\tau \in [0, t_B]$ and sufficiently large x, which means that the semigroup $\{P(t)\}_{t \geq 0}$ is weakly tight. Summarizing, if $g(2\bar{x}) \neq 2g(\bar{x})$ for some $\bar{x} > x_{\min}$ and

$$
\limsup_{x \to \infty} \frac{g(x)}{x\varphi(x)} < \log 2 - t_B \limsup_{x \to \infty} \frac{g(x)}{x},
$$

then, according to Theorem 5.6, the semigroup $\{P(t)\}_{t \geq 0}$ is asymptotically stable. Moreover, if f^* is the invariant density, then $f^*(x, 1) > 0 \iff x > x_{\min}$ and $f^*(x, \tau, 2) > 0 \iff x > \pi_\tau x_B$.

6.3.5 Lebowitz–Rubinow's Model

We recall that in this model a maturity of a cell is a real variable $x \in [0, 1]$. A new born cell has maturity 0 and a cell splits at maturity 1. The maturity x grows according to the equation $x' = v$, where the maturation velocity v of each individual cell is constant. The relation between the maturation velocities of mother's v and daughter's cells v' is given by a transition probability $P(v, dv')$. As we have mentioned in Sect. 1.7, this model is a special case of one-dimensional stochastic billiards investigated in [78] and based on that paper we briefly present results concerning its asymptotic properties. We assume that $v \in (0, 1]$ and $P(v, dv') = \alpha \delta_v(v') + \beta k(v', v) \, dv'$, where $\beta \in (0, 1]$, $\alpha + \beta = 1$, and $\int_0^1 k(v', v) \, dv' = 1$ for $v \in (0, 1]$. We showed in Sect. 4.2.6 that the process $\xi(t) = (x(t), v(t))$ induces a stochastic semigroup $\{P(t)\}_{t \geq 0}$ on the space $L^1(X, \Sigma, m)$, where $X = (0, 1]^2$, $\Sigma = \mathcal{B}(X)$ and $m(dx \times dv) = dx \times dv$. Since the boundary condition (4.26) contains a kernel operator, one can check that the semigroup $\{P(t)\}_{t \geq 0}$ is partially integral. Observe that if this semigroup has an invariant density f^*, then $Af^* = 0$ and from (4.25) it follows that f^* does not depend on x. From (4.26) we deduce that $v f^* = K(v f^*)$, where K is a stochastic operator on $L^1[0, 1]$ given by

$$Kh(v') = \int_0^1 k(v', v) h(v) \, dv. \tag{6.39}$$

We assume that K is irreducible which is equivalent to say that there does not exist a set $B \subseteq (0, 1]$ of the Lebesgue measure $0 < |B| < 1$ such that $k(v', v) = 0$ for $v \in B$ and $v' \notin B$. From irreducibility it follows that if an invariant density f^* exists then it is unique and $f^*(x) > 0$ for a.e. x. The question of the existence of an invariant density is nontrivial. If for example we assume that there exist $C > 0$ and $\gamma > 0$ such that

$$k(v, v') \leq C|v|^\gamma \quad \text{for } v, v' \in (0, 1], \tag{6.40}$$

then an invariant density exists. It means that irreducibility of K and condition (6.40) implies asymptotic stability of the semigroup. Now we consider the case when there is no invariant density. Assume that the kernel k is bounded and K is irreducible. Then the operator K has a unique invariant density $h^* > 0$ and the function $f^*(v) = v^{-1} h^*(v)$ is invariant with respect to the semigroup $\{P(t)\}_{t \geq 0}$. Since f^* is not integrable, but $\int_\varepsilon^1 f^*(v) \, dv < \infty$ for every $\varepsilon > 0$, from Theorem 5.4(c) it follows that

$$\lim_{t \to \infty} \int_0^\varepsilon \int_0^1 P(t) f(x, v) \, dx \, dv = 1 \tag{6.41}$$

for every density f and every $\varepsilon > 0$. For example if $k \equiv 1$, then $h^* \equiv 1$ and $f^*(v) = v^{-1}$ is not integrable. Thus, the semigroup has no invariant density and consequently (6.41) holds. It is interesting that in this example we have

$$P(t) f(x, v) \sim \frac{c}{|v|} (\log t)^{-1} \quad \text{as } t \to \infty$$

for $v \geq \varepsilon$ and $x \in [0, 1]$, where c is some constant.

6.3.6 Stein's Model

In Sect. 1.11 we considered Stein's model which describes how the depolarization $V(t)$ changes in time. Now, based on the paper [91], we present its asymptotic properties. First, we introduce an extra 0-phase, which begins at the end of the refractory period and ends when depolarization jumps from 0 for the first time. Thus the process is defined on the space $X = \{(0, 0)\} \cup (-\infty, \theta] \times \{1\} \cup [0, t_R] \times \{2\}$. Define a measure m on the σ-algebra $\mathscr{B}(X)$ of the Borel subsets of X by $m = \delta_{(0,0)} + m_1 + m_2$, where $\delta_{(0,0)}$ is the Dirac measure at $(0, 0)$, m_1 is the Lebesgue measure on the segment $(-\infty, \theta] \times \{1\}$ and m_2 is the Lebesgue measure on the segment $[0, t_R] \times \{2\}$. The process induces a stochastic semigroup $\{P(t)\}_{t \geq 0}$ on the space $L^1(X, \mathscr{B}(X), m)$.

The stochastic semigroup introduced by Stein's model is asymptotically stable. The proof of this result is given in [91] and it is based on Theorem 5.6. Since the semigroup is asymptotically stable it would be useful to find an invariant density. Unfortunately, this is rather a difficult task and we only present the fact that the invariant density f^* of the process with the refractory period can be expressed by the invariant density $\bar{f}^*$ of the process without the refractory period (i.e. $t_R = 0$) and the same other parameters. Let

$$c = \left(1 + t_R(\lambda_E + \lambda_I)\bar{f}^*(0, 0)\right)^{-1}.$$

Then

$$f^*(0, 0) = c\bar{f}^*(0, 0),$$
$$f^*(x, 1) = c\bar{f}^*(x, 1) \quad \text{for } x \in (-\infty, \theta],$$
$$f^*(x, 2) = c(\lambda_E + \lambda_I)\bar{f}^*(0, 0) \quad \text{for } x \in [0, t_R].$$

Appendix A
Measure and Probability Essentials

Our objective in this appendix is to introduce basic definitions and results from measure and probability theory required in the text. For the general background and proofs we refer the reader to [55].

A. 1 Measurable Spaces and Mappings

A *measurable space* is a pair (X, Σ) consisting of a non-empty set X and a family Σ of subsets of X, called a *σ-algebra*, satisfying the following

(1) $X \in \Sigma$,
(2) if $B \in \Sigma$ then its complement $B^c := X \setminus B$ is in Σ,
(3) if $B_1, B_2, \ldots \in \Sigma$ then $\bigcup_n B_n \in \Sigma$.

Let $\mathscr{C}$ be a family of subsets of X. Then there exists a smallest σ-algebra containing all sets from $\mathscr{C}$. It is the intersection of all σ-algebras of subsets of X containing the family $\mathscr{C}$. We denote it by $\sigma(\mathscr{C})$ and we call it the *σ-algebra generated by $\mathscr{C}$*. Note that if X is a metric space, then the Borel σ-algebra $\mathscr{B}(X)$ is the σ-algebra generated by all open subsets of X and its elements are called *Borel sets*.

Let (X_1, Σ_1) and (X_2, Σ_2) be measurable spaces. For $B_1 \in \Sigma_1$ and $B_2 \in \Sigma_2$ we write $B_1 \times B_2$ for the set of all pairs (x_1, x_2) with $x_1 \in B_1$ and $x_2 \in B_2$ and call it a *measurable rectangle* in $X_1 \times X_2$. We let $\Sigma_1 \otimes \Sigma_2$ to be the σ-algebra generated by the family of all measurable rectangles, so that

$$\Sigma_1 \otimes \Sigma_2 = \sigma(\mathscr{C}), \quad \text{where } \mathscr{C} = \{B_1 \times B_2 \colon B_1 \in \Sigma_1, B_2 \in \Sigma_2\}.$$

Note that the intersection of two measurable rectangles is a measurable rectangle. The σ-algebra $\Sigma_1 \otimes \Sigma_2$ is called the *product σ-algebra* on $X_1 \times X_2$, and the measurable space $(X_1 \times X_2, \Sigma_1 \otimes \Sigma_2)$ is called the *product* of (X_1, Σ_1) and (X_2, Σ_2).

© The Author(s) 2017
R. Rudnicki and M. Tyran-Kamińska, *Piecewise Deterministic Processes in Biological Models*, SpringerBriefs in Mathematical Methods,
DOI 10.1007/978-3-319-61295-9

A family $\mathscr{C}$ of subsets of X is called a π-*system* if it is closed under intersections, i.e. if $B_1, B_2 \in \mathscr{C}$ then $B_1 \cap B_2 \in \mathscr{C}$. A family $\mathscr{B}$ of subsets of X is a λ-*system* if:

(1) $X \in \mathscr{B}$.
(2) If $B_1, B_2 \in \mathscr{B}$ and $B_1 \subset B_2$ then $B_2 \setminus B_1 \in \mathscr{B}$.
(3) If $B_1, B_2, \ldots \in \mathscr{B}$ and $B_n \subset B_{n+1}$ then $\bigcup_n B_n \in \mathscr{B}$.

Note that a family of subsets of X is a σ-algebra if and only if it is a π-system and λ-system. We have the fundamental result for establishing measurability.

Lemma A.1 $(\pi - \lambda$ lemma) *If $\mathscr{C}$ is a π-system and $\mathscr{B}$ is a λ-system such that $\mathscr{C} \subset \mathscr{B}$ then $\sigma(\mathscr{C}) \subset \mathscr{B}$.*

A mapping $f : X_1 \to X_2$ is said to be *measurable* if the counter image of any $B \in \Sigma_2$ under f is a measurable subset of X_1, i.e.

$$f^{-1}(B) = \{x \in X_1 : f(x) \in B\} \in \Sigma_1.$$

If $X_2 = \mathbb{R}$ then we take $\Sigma_2 = \mathscr{B}(\mathbb{R})$. If we want to underline the σ-algebra in the *domain* X_1 then we say that f is *measurable with respect to* Σ_1, or Σ_1-*measurable*.

A measurable function $f : X \to \mathbb{R}$ is called a *simple function* if there exist a finite number of pairwise disjoint sets $B_1, \ldots, B_k \in \Sigma$ and constants $c_1, \ldots, c_k$ such that

$$f(x) = \sum_i c_i \mathbf{1}_{B_i}(x), \quad x \in X.$$

Observe that every non-negative measurable function $f : X \to [0, \infty]$ is a pointwise limit of a monotone sequence of simple functions, since

$$f(x) = \lim_{n \to \infty} f_n(x), \quad x \in X,$$

where

$$f_n(x) = \begin{cases} (k-1)2^{-n}, & \text{if } (k-1)2^{-n} \le f(x) < k2^{-n}, \ 1 \le k \le n2^n, \\ n, & \text{if } n \le f(x). \end{cases}$$

Finally, if f is measurable then $f = f^+ - f^-$, where $f^+ = \max\{0, f\}$ and $f^- = \max\{0, -f\}$ are, respectively, the positive part and the negative part of f.

Many results concerning measurable functions are usually easily proved for indicator functions and we can extend them to all measurable functions by using the monotone class argument based on the following result.

Theorem A.1 (Monotone class theorem) *Let $\mathscr{C}$ be a π-system of subsets of X such that $X \in \mathscr{C}$. Let $\mathscr{H}$ be a class of real-valued functions on X such that*

(1) *If $B \in \mathscr{C}$ then $\mathbf{1}_B \in \mathscr{H}$.*
(2) *If $f, g \in \mathscr{H}$ then $f + g \in \mathscr{H}$ and $cf \in \mathscr{H}$ for any non-negative (resp. real) constant c.*

(3) *If $f_n \in \mathcal{H}$ are non-negative and $f_n \uparrow f$ with a non-negative (and bounded) f then $f \in \mathcal{H}$.*

Then $\mathcal{H}$ contains all non-negative (bounded) functions measurable with respect to $\sigma(\mathcal{C})$.

A. 2 Measure and Integration

A *measure* on (X, Σ) is a mapping $\mu\colon \Sigma \to [0, \infty]$ such that $\mu(\emptyset) = 0$ and μ is *countably additive*, i.e.

$$\mu(\bigcup_n B_n) = \sum_n \mu(B_n)$$

for every sequence of disjoint sets $B_n \in \Sigma$. A measure μ is called a *probability measure* if $\mu(X) = 1$, a *finite measure* if $\mu(X) < \infty$, and a *σ-finite measure* if there is a sequence $B_j \in \Sigma$ such that $X = \bigcup_j B_j$ and $\mu(B_j) < \infty$, for each j. The triple (X, Σ, μ) is said to be a *(probability, finite, σ-finite) measure space* if μ is a (probability, finite, σ-finite) measure on (X, Σ).

Theorem A.2 (Uniqueness) *Let μ and ν be two finite measures on (X, Σ). Suppose that $\mathcal{C}$ is a π-system such that $X \in \mathcal{C}$ and $\sigma(\mathcal{C}) = \Sigma$. If $\mu(B) = \nu(B)$ for all $B \in \mathcal{C}$ then $\mu = \nu$.*

Let (X, Σ, μ) be a measure space. The *Lebesgue integral* of a simple function $f = \sum_i c_i \mathbf{1}_{B_i}$ is defined as

$$\int_X f(x)\,\mu(dx) = \sum_i c_i \mu(B_i)$$

and that of a non-negative measurable function f as

$$\int_X f(x)\,\mu(dx) = \lim_{n\to\infty} \int_X f_n(x)\,\mu(dx),$$

where (f_n) is a non-decreasing sequence of simple measurable functions converging pointwise to f. Finally, if f is measurable then we set

$$\int_X f(x)\,\mu(dx) = \int_X f^+(x)\,\mu(dx) - \int_X f^-(x)\,\mu(dx)$$

if at least one of the terms

$$\int_X f^+(x)\,\mu(dx), \quad \int_X f^-(x)\,\mu(dx)$$

is finite. If both are finite then f is said to be μ-*integrable*, or simply *integrable*.

Lemma A.2 (Fatou) *Let (X, Σ, μ) be a measure space and let (f_n) be a sequence of non-negative measurable functions. Then*

$$\int_X \liminf_{n \to \infty} f_n(x) \, \mu(dx) \le \liminf_{n \to \infty} \int_X f_n(x) \, \mu(dx).$$

Theorem A.3 (Lebesgue) *Let (X, Σ, μ) be a measure space and let (f_n) be a sequence of measurable functions.*

(1) Monotone convergence: *If $0 \le f_n(x) \le f_{n+1}(x)$ for all n and x then*

$$\lim_{n \to \infty} \int_X f_n(x) \, \mu(dx) = \int_X \lim_{n \to \infty} f_n(x) \, \mu(dx). \tag{A.1}$$

(2) Dominated convergence: *If g is integrable, $|f_n(x)| \le g(x)$ for all n and x, and the sequence (f_n) is convergent pointwise, then (A.1) holds.*

The next theorem, referred to usually as the Fubini theorem, is about the existence of a product measure and the change of the order of integration in double integrals.

Theorem A.4 (Fubini) *Let (X_1, Σ_1, μ_1) and (X_2, Σ_2, μ_2) be σ-finite measure spaces. Then there exists a unique measure $\mu_1 \times \mu_2$ on $(X_1 \times X_2, \Sigma_1 \otimes \Sigma_2)$ satisfying*

$$(\mu_1 \times \mu_2)(B_1 \times B_2) = \mu_1(B_1)\mu_2(B_2), \quad B_1 \in \Sigma_1, B_2 \in \Sigma_2.$$

Moreover, if $f : X_1 \times X_2 \to [0, \infty]$ is $\Sigma_1 \otimes \Sigma_2$-measurable then

$$\int_{X_1 \times X_2} f(x_1, x_2) \, \mu_1 \times \mu_2(dx_1, dx_2) = \int_{X_1} \left(\int_{X_2} f(x_1, x_2) \, \mu_2(dx_2) \right) \mu_1(dx_1)$$
$$= \int_{X_2} \left(\int_{X_1} f(x_1, x_2) \, \mu_1(dx_1) \right) \mu_2(dx_2).$$

The result remains true for any $\mu_1 \times \mu_2$-integrable f. The measure $\mu_1 \times \mu_2$ is called the product measure *of μ_1 and μ_2.*

For a given measure space (X, Σ, μ) a finite signed measure ν on (X, Σ) is called *absolutely continuous with respect to* μ if $\nu(B) = 0$ whenever $\mu(B) = 0$. In this case ν is also said to be *dominated by* μ, written $\nu \ll \mu$.

Theorem A.5 (Radon–Nikodym) *Let (X, Σ, μ) be a σ-finite measure space and let ν be a finite signed measure on (X, Σ) absolutely continuous with respect to μ. Then there exists $f \in L^1(X, \Sigma, \mu)$ such that*

$$\nu(B) = \int_B f(x) \, \mu(dx) \quad \text{for all } B \in \Sigma.$$

The function f is unique as an element of $L^1(X, \Sigma, \mu)$ and it is called the Radon–Nikodym derivative $\dfrac{d\nu}{d\mu}$ *of ν with respect to μ.*

A. 3 Random Variables

Let $(\Omega, \mathscr{F}, \mathbb{P})$ be a probability space and let (X, Σ) be a measurable space. An X-valued random variable ξ is called *discrete*, if X is a finite or a countable set. If $X = \mathbb{R}^d$ then we say that a random variable is *continuous* if its distribution is absolutely continuous with respect to the Lebesgue measure. Then the Radon–Nikodym theorem implies that the distribution of ξ has a *density*, i.e. there exists a Borel measurable non-negative function f_ξ such that

$$\mu_\xi(B) = \mathbb{P}(\xi \in B) = \int_B f_\xi(x)\, dx, \quad B \in \mathscr{B}(X).$$

Given a real-valued random variable ξ the function

$$F(t) = \mathbb{P}(\xi \le t), \quad t \in \mathbb{R},$$

is called the *distribution function* of the random variable ξ. It is non-decreasing, right-continuous, and

$$\lim_{t \to -\infty} F(t) = 0, \quad \lim_{t \to \infty} F(t) = 1.$$

A function $F \colon \mathbb{R} \to \mathbb{R}$ is called *absolutely continuous* if and only if for every $\varepsilon > 0$ there exists a $\delta > 0$ such that for each finite collection of disjoint intervals $(a_1, b_1), \ldots, (a_k, b_k)$

$$\sum_{j=1}^k |F(b_j) - F(a_j)| < \varepsilon \quad \text{if} \quad \sum_{j=1}^k (b_j - a_j) < \delta.$$

In particular, a distribution function of a real-valued random variable is absolutely continuous if and only if the random variable has a density.

The expectation of a real-valued random variable ζ is defined as the integral of ζ with respect to $\mathbb{P}$

$$\mathbb{E}(\zeta) = \int_\Omega \zeta(\omega)\mathbb{P}(d\omega) = \int \zeta\, d\mathbb{P}.$$

If $\zeta = h(\xi)$ with measurable h and if the X-valued random variable ξ has distribution μ_ξ then

$$\mathbb{E}(h(\xi)) = \int_X h(x)\,\mu_\xi(dx).$$

In particular, if the random variable ξ is discrete then

$$\mathbb{E}(h(\xi)) = \sum_{x \in X} h(x)\mathbb{P}(\xi = x)$$

and if ξ is continuous with density f_ξ then

$$\mathbb{E}(h(\xi)) = \int_X h(x)f_\xi(x)\,dx.$$

Families $\mathscr{F}_1, \ldots, \mathscr{F}_n \subset \mathscr{F}$ of events are said to be *independent* if for any sets $A_1 \in \mathscr{F}_1, \ldots, A_n \in \mathscr{F}_n$ we have

$$\mathbb{P}(A_1 \cap \ldots \cap A_n) = \mathbb{P}(A_1)\ldots\mathbb{P}(A_n).$$

We say that $\xi_1, \ldots, \xi_n$ are *independent random variables* if the corresponding σ-algebras $\sigma(\xi_1), \ldots, \sigma(\xi_n)$ are independent. A collection of random variables is said to be independent if each finite number of them is independent. We have the following characterization.

Theorem A.6 *Let $\xi_1, \ldots, \xi_n$ be random variables with values in some measurable spaces $X_1, \ldots, X_n$ and have distributions $\mu_1, \ldots, \mu_n$. Then $\xi_1, \ldots, \xi_n$ are independent if and only if the distribution of $(\xi_1, \ldots, \xi_n)$ is the product measure $\mu_1 \times \cdots \times \mu_n$.*

Theorem A.7 (Strong law of large numbers) *Let $\xi_1, \xi_2, \ldots$ be i.i.d. random variables such that $\mathbb{E}(|\xi_1|) < \infty$. Then*

$$\mathbb{P}\left(\lim_{n \to \infty} \frac{1}{n} \sum_{k=1}^n \xi_k = \mathbb{E}(\xi_1)\right) = 1.$$

A. 4 Conditional Expectations and Distributions

Let $(\Omega, \mathscr{F}, \mathbb{P})$ be a probability space. Suppose that $\zeta \colon \Omega \to \mathbb{R}$ is an integrable random variable, i.e. $\mathbb{E}|\zeta| < \infty$. The *conditional expectation of ζ given an event* $A \in \mathscr{F}$ with $\mathbb{P}(A) > 0$ is defined as the number

$$\mathbb{E}(\zeta|A) = \int_\Omega \zeta(\omega)\mathbb{P}(d\omega|A) = \frac{1}{\mathbb{P}(A)} \int_A \zeta(\omega)\mathbb{P}(d\omega),$$

where $\mathbb{P}(\cdot|A) \colon \mathscr{F} \to [0, 1]$ is the probability measure defined by

$$\mathbb{P}(B|A) = \frac{\mathbb{P}(A \cap B)}{\mathbb{P}(A)}, \quad B \in \mathscr{F}.$$

Given a σ-algebra $\mathscr{G} \subset \mathscr{F}$ the *conditional expectation* of ζ with respect to $\mathscr{G}$ is a real-valued random variable η such that η is $\mathscr{G}$-measurable and

$$\int_A \zeta(\omega)\mathbb{P}(d\omega) = \int_A \eta(\omega)\mathbb{P}(d\omega) \quad \text{for all } A \in \mathscr{G}. \tag{A.2}$$

It is denoted by $\mathbb{E}(\zeta|\mathscr{G})$.

Theorem A.8 (Existence) *If ζ is an integrable random variable then the conditional expectation of ζ with respect to a σ-algebra $\mathscr{G}$ exists and is unique, as an element of $L^1(\Omega, \mathscr{G}, \mathbb{P})$.*

Theorem A.9 (Properties of conditional expectation) *The following properties hold whenever the corresponding expressions exist for the absolute values:*

(1) $\mathbb{E}(\mathbb{E}(\zeta|\mathscr{G})) = \mathbb{E}(\zeta)$.
(2) $\mathbb{E}(\zeta_1 + \zeta_2|\mathscr{G}) = \mathbb{E}(\zeta_1|\mathscr{G}) + \mathbb{E}(\zeta_2|\mathscr{G})$ *and* $\mathbb{E}(c\zeta|\mathscr{G}) = c\mathbb{E}(\zeta|\mathscr{G})$.
(3) *If* $\zeta \geq 0$ *then* $\mathbb{E}(\zeta|\mathscr{G}) \geq 0$.
(4) *If* $\zeta_n \uparrow \zeta$ *then* $\mathbb{E}(\zeta_n|\mathscr{G}) \uparrow \mathbb{E}(\zeta|\mathscr{G})$.
(5) *If* ξ *is* $\mathscr{G}$-*measurable then* $\mathbb{E}(\xi\zeta|\mathscr{G}) = \xi\mathbb{E}(\zeta|\mathscr{G})$.
(6) *If* $\mathscr{G} \subset \mathscr{H}$ *then* $\mathbb{E}(\mathbb{E}(\zeta|\mathscr{H})|\mathscr{G}) = \mathbb{E}(\zeta|\mathscr{G})$.

The *conditional probability* of an event $A \in \mathscr{F}$, given a σ-algebra $\mathscr{G}$, is defined as

$$\mathbb{P}(A|\mathscr{G}) = \mathbb{E}(\mathbf{1}_A|\mathscr{G}).$$

If $\mathscr{G} = \sigma(\xi)$ then we write $\mathbb{E}(\zeta|\xi)$ to denote $\mathbb{E}(\zeta|\sigma(\xi))$, thus we also have

$$\mathbb{P}(A|\xi) = \mathbb{E}(\mathbf{1}_A|\sigma(\xi)), \quad A \in \mathscr{F}.$$

In particular, the random variable $\mathbb{P}(\zeta \in B|\xi)$, being $\sigma(\xi)$ measurable, can be represented as $h_B(\xi)$ for some measurable h_B, since we have the following.

Theorem A.10 (Representation theorem) *Let ξ be an X-valued random variable. Then a real-valued random variable η is measurable with respect to $\sigma(\xi)$ if and only if there exists a measurable $h\colon X \to \mathbb{R}$ such that $\eta = h(\xi)$.*

The next result is useful for the computation of conditional expectations.

Lemma A.3 *Let ξ be a $\mathscr{G}$-measurable X-valued random variable and ζ be independent of $\mathscr{G}$. If g is a measurable non-negative function and $h(x) = \mathbb{E}(g(x, \zeta))$, $x \in X$, then*

$$\mathbb{E}(g(\xi, \zeta)|\mathscr{G}) = h(\xi). \tag{A.3}$$

Proof Since ξ is $\mathscr{G}$-measurable, the random variable $h(\xi)$ is $\mathscr{G}$-measurable. It remains to show that

$$\int_A g(\xi(\omega), \zeta(\omega)) \, \mathbb{P}(d\omega) = \int_A h(\xi(\omega)) \, \mathbb{P}(d\omega)$$

for every $A \in \mathscr{G}$, or, equivalently, $\mathbb{E}(\eta g(\xi, \zeta)) = \mathbb{E}(\eta h(\xi))$ for every positive $\mathscr{G}$-measurable random variable η. Let μ_ζ and $\mu_{(\xi,\eta)}$ denote the distributions of ζ and (ξ, η). We have

$$h(x) = \mathbb{E}(g(x, \zeta)) = \int g(x, y) \, \mu_\zeta(dy).$$

Since (ξ, η) and ζ are independent, we obtain

$$\mathbb{E}(\eta g(\xi, \zeta)) = \int zg(x, y) \, \mu_{(\xi,\eta)}(dx, dz) \, \mu_\zeta(dy)$$

$$= \int z \left(\int g(x, y) \, \mu_\zeta(dy) \right) \mu_{(\xi,\eta)}(dx, dz),$$

which implies that

$$\mathbb{E}(\eta g(\xi, \zeta)) = \int zh(x) \, \mu_{(\xi,\eta)}(dx, dz) = \mathbb{E}(\eta h(\xi))$$

and completes the proof.

Let $\mathscr{K}, \mathscr{G}, \mathscr{H} \subseteq \mathscr{F}$ be σ-algebras such that $\mathscr{H} \subseteq \mathscr{G} \cap \mathscr{K}$. We say that $\mathscr{G}$ and $\mathscr{K}$ are *conditionally independent given* $\mathscr{H}$, if

$$\mathbb{P}(A \cap B | \mathscr{H}) = \mathbb{P}(A | \mathscr{H}) \mathbb{P}(B | \mathscr{H}), \quad A \in \mathscr{G}, \ B \in \mathscr{K}.$$

Lemma A.4 *Let $\mathscr{K}, \mathscr{G}, \mathscr{H} \subseteq \mathscr{F}$ be σ-algebras such that $\mathscr{H} \subseteq \mathscr{G} \cap \mathscr{K}$. Then the following are equivalent:*

(1) $\mathbb{P}(A \cap B | \mathscr{H}) = \mathbb{P}(A | \mathscr{H}) \mathbb{P}(B | \mathscr{H})$ *for all $A \in \mathscr{G}, B \in \mathscr{K}$.*
(2) $\mathbb{P}(A | \mathscr{K}) = \mathbb{P}(A | \mathscr{H})$ *for all $A \in \mathscr{G}$.*

Proof We show that (1) implies (2). Since for any A the random variable $\mathbb{P}(A | \mathscr{H})$ is $\mathscr{K}$-measurable, it is enough to show that for every $B \in \mathscr{K}$ we have

$$\mathbb{E}(\mathbf{1}_B \mathbb{E}(\mathbf{1}_A | \mathscr{K})) = \mathbb{E}(\mathbf{1}_B \mathbb{E}(\mathbf{1}_A | \mathscr{H})).$$

Since $\mathbf{1}_B$ is $\mathscr{K}$-measurable, we obtain

$$\mathbb{E}(\mathbf{1}_B \mathbb{E}(\mathbf{1}_A | \mathscr{K})) = \mathbb{E}(\mathbb{E}(\mathbf{1}_B \mathbf{1}_A | \mathscr{K})) = \mathbb{E}(\mathbf{1}_B \mathbf{1}_A) = \mathbb{P}(A \cap B).$$

On the other hand

$$\mathbb{E}(\mathbf{1}_B\mathbb{E}(\mathbf{1}_A|\mathcal{H})) = \mathbb{E}(\mathbb{E}(\mathbf{1}_B\mathbb{E}(\mathbf{1}_A|\mathcal{H})|\mathcal{H})) = \mathbb{E}(\mathbb{E}(\mathbf{1}_A|\mathcal{H})\mathbb{E}(\mathbf{1}_B|\mathcal{H}))$$

which, by assumption, we can rewrite as

$$\mathbb{E}(\mathbb{E}(\mathbf{1}_B|\mathcal{H})\mathbb{E}(\mathbf{1}_A|\mathcal{H})) = \mathbb{E}(\mathbb{E}(\mathbf{1}_{B\cap A}|\mathcal{H})) = \mathbb{P}(A \cap B).$$

Finally, to show that (2) implies (1), we observe that

$$\mathbb{P}(B \cap A|\mathcal{H}) = \mathbb{E}(\mathbf{1}_{B\cap A}|\mathcal{H}) = \mathbb{E}(\mathbb{E}(\mathbf{1}_{B\cap A}|\mathcal{K})|\mathcal{H}) = \mathbb{E}(\mathbf{1}_B\mathbb{E}(\mathbf{1}_A|\mathcal{K})|\mathcal{H})$$
$$= \mathbb{E}(\mathbf{1}_B\mathbb{E}(\mathbf{1}_A|\mathcal{H})|\mathcal{H}) = \mathbb{E}(\mathbf{1}_A|\mathcal{H})\mathbb{E}(\mathbf{1}_B|\mathcal{H}).$$

References

1. Allen, L.J.S.: An Introduction to Stochastic Processes with Applications to Biology. Chapman Hall/CRC Press, Boca Raton (2010)
2. Almeida, C.R., de Abreu, F.V.: Dynamical instabilities lead to sympatric speciation. Evol. Ecol. Res. **5**, 739–757 (2003)
3. Alur R., Pappas, G.J. (eds.): Hybrid Systems: Computation and Control. LNCS, vol. 2993. Springer, Berlin (2004)
4. Arendt, W.: Resolvent positive operators. Proc. Lond. Math. Soc. **54**, 321–349 (1987)
5. Arendt, W., Grabosch, A., Greiner, G., Groh, U., Lotz, H.P., Moustakas, U., Nagel, R., Neubrander, F., Schlotterbeck, U.: One-parameter semigroups of positive operators. Lecture Notes in Mathematics, vol. 1184. Springer, Berlin (1986)
6. Arino, O., Rudnicki, R.: Phytoplankton dynamics. C. R. Biologies **327**, 961–969 (2004)
7. Bachar, M., Batzel, J., Ditlevsen, S. (eds.): Stochastic Biomathematical Models with Applications to Neuronal Modeling. Lecture Notes in Mathematics, vol 2058, Springer, Berlin, Heidelberg (2013)
8. Bakhtin, Y., Hurth, T.: Invariant densities for dynamical system with random switching. Nonlinearity **25**, 2937–2952 (2012)
9. Banasiak, J.: On an extension of the Kato-Voigt perturbation theorem for substochastic semigroups and its application. Taiwanese J. Math. **5**, 169–191 (2001)
10. Banasiak, J., Arlotti, L.: Perturbations of positive semigroups with applications. London Ltd., Springer Monographs in Mathematics. Springer, London (2006)
11. Banasiak, J., Pichór, K., Rudnicki, R.: Asynchronous exponential growth of a general structured population model. Acta Applicandae Mathematicae **119**, 149–166 (2012)
12. Bell, G.I., Anderson, E.C.: Cell growth and division I. A Mathematical model with applications to cell volume distributions in mammalian suspension cultures. Biophys. J. **7**, 329–351 (1967)
13. Benaïm, M., le Borgne, S., Malrieu, F., Zitt, P.A.: Qualitative properties of certain piecewise deterministic Markov processes. Ann. Inst. H. Poincaré Probab. Stat. **51**, 1040–1075 (2015)
14. Bernoulli, D., Blower, S.: An attempt at a new analysis of the mortality caused by smallpox and of the advantages of inoculation to prevent it. Rev. Med. Virol. **14**, 275–288 (2004)
15. Biedrzycka, W., Tyran-Kamińska, M.: Existence of invariant densities for semiflows with jumps. J. Math. Anal. Appl. **435**, 61–84 (2016)
16. Blom, H.A.P., Lygeros, J. (eds.): Stochastic Hybrid Systems: Theory and Safety Critical Applications. Series: Lecture Notes in Control and Information Sciences, vol. 337. Springer, Berlin Heidelberg (2006)

© The Author(s) 2017

R. Rudnicki and M. Tyran-Kamińska, *Piecewise Deterministic Processes in Biological Models*, SpringerBriefs in Mathematical Methods, DOI 10.1007/978-3-319-61295-9

17. Bobrowski, A., Lipniacki, T., Pichór, K., Rudnicki, R.: Asymptotic behavior of distributions of mRNA and protein levels in a model of stochastic gene expression. J. Math. Anal. Appl. **333**, 753–769 (2007)
18. Bokes, P., King, J.R., Wood, A.T.A., Loose, M.: Transcriptional bursting diversifies the behaviour of a toggle switch: Hybrid simulation of stochastic gene expression. Bull. Math. Biol. **75**, 351–371 (2013)
19. Boulanouar, M.: A mathematical study for a Rotenberg model. J. Math. Anal. Appl. **265**, 371–394 (2002)
20. Bressloff, P., Newby, J.: Directed intermittent search for hidden targets. New J. Phys. **11**, 023033 (2009)
21. Bressloff, P., Newby, J.: Quasi-steady state reduction of molecular motor-based models of directed intermittent search. Bull. Math. Biol. **72**, 1840–1866 (2010) Rev. Mod. Phys. **85**, 135–196 (2013)
22. Bressloff, P., Newby, J.: Stochastic models of intracellular transport. Rev. Mod. Phys. **85**, 135–196 (2013)
23. Buckwar, E., Riedler, M.G.: An exact stochastic hybrid model of excitable membranes including spatio-temporal evolution. J. Math. Biol. **63**, 1051–1093 (2011)
24. Burkitt, A.N.: A review of the integrate-and-fire neuron model: I Homogeneous synaptic input. Biol. Cybern. **95**, 1–19 (2006)
25. Cai, L., Friedman, N., Xie, X.S.: Stochastic protein expression in individual cells at the single molecule level. Nature **440**, 358–362 (2006)
26. Capasso, V., Bakstein, D.: An Introduction to Continuous-Time Stochastic Processes. Theory, Models and Applications to Finance, Biology and Medicine. Birkhäuser, Boston (2005)
27. Cassandras, C.G., Lygeros, J. (eds.): Stochastic Hybrid Systems. Control Engineering Series, vol. 24, CRC Press, Boca Raton, FL (2007)
28. Davis, M.H.A.: Piecewise-deterministic Markov processes: a general class of nondiffusion stochastic models. J. R. Stat. Soc. Ser. B **46**, 353–388 (1984)
29. Davis, M.H.A.: Markov Models and Optimization. Monographs on Statistics and Applied Probability, vol. 49, Chapman & Hall, London (1993)
30. Desch, W.: Perturbations of positive semigroups in AL-spaces. Unpublished (1988)
31. Diekmann, O., Heijmans, H.J.A.M., Thieme, H.R.: On the stability of the cell size distribution. J. Math. Biol. **19**, 227–248 (1984)
32. Dieudonne, J.: Sur le théorème de Radon-Nikodym. Ann. Univ. Grenoble **23**, 25–53 (1948)
33. Dunford, N.: On one parameter groups of linear transformations. Ann. Math. **39**, 569–573 (1938)
34. Engel, K.J., Nagel, R.: One-parameter semigroups for linear evolution equations. Graduate Texts in Mathematics, vol. 194. Springer, New York (2000)
35. Ethier, S.N., Kurtz, T.G.: Markov Processes. Characterization and Convergence. Wiley, New York (1986)
36. Evans, S.N.: Stochastic billiards on general tables. Ann. Appl. Probab. **11**, 419–437 (2001)
37. Feller, W.: An introduction to Probability Theory and Applications. Wiley, New York (1966)
38. Foguel, S.R.: The Ergodic Theory of Markov Processes. Van Nostrand Reinhold Comp, New York (1969)
39. Fox, R.F.: Stochastic versions of the Hodgkin-Huxley equations. Biophys. J. **72**, 2068–2074 (1997)
40. Friedman, N., Cai, L., Xie, X.S.: Linking stochastic dynamics to population distribution: an analytical framework of gene expression. Phys. Rev. Lett. **97**, 168302 (2006)
41. Gavrilets, S., Boake, C.R.B.: On the evolution of premating isolation after a founder event. Am. Nat. **152**, 706–716 (1998)
42. Genadot, A., Thieullen, M.: Averaging for a fully coupled piecewise-deterministic Markov process in infinite dimensions. Adv. Appl. Probab. **44**, 749–773 (2012)
43. Golding, I., Paulsson, J., Zawilski, S., Cox, E.: Real-time kinetics of gene activity in individual bacteria. Cell **123**, 1025–1036 (2005)

44. Greiner, G.: Perturbing the boundary conditions of a generator. Houston J. Math. **13**, 213–229 (1987)
45. Gwiżdż, P., Tyran-Kamińska, M.: Densities for piecewise deterministic Markov processes with boundary. Preprint
46. Gyllenberg, M., Heijmans, H.J.A.M.: An abstract delay-differential equation modelling size dependent cell growth and division. SIAM J. Math. Anal. **18**, 74–88 (1987)
47. Hasminskiǐ, R.Z.: Ergodic properties of recurrent diffusion processes and stabilization of the solutions of the Cauchy problem for parabolic equations (in Russian). Teor. Verojatn. Primenen. **5**, 196–214 (1960)
48. Hennequin, P., Tortrat, A.: Théorie des probabilités et quelques applications. Masson et Cie, Paris (1965)
49. Hille, E.: Functional Analysis and Semi-Groups. American Mathematical Society Colloquium Publications, vol. 31. American Mathematical Society, New York (1948)
50. Hillen, T., Hadeler, K.P.: Hyperbolic systems and transport equations in mathematical biology. In: Warnecke, G. (ed.) Analysis and Numerics for Conservation Laws, pp. 257–279. Springer, Berlin, Heidelberg (2005)
51. Hu, J., Lygeros, J. Sastry, S.: Towards a theory of stochastic hybrid systems. In: Lynch, N., Krogh, B,H. (eds.) Hybrid Systems: Computation and Control. LNCS vol. 1790, pp. 160–173. Springer, Berlin (2000)
52. Hu, J., Wu, W.C., Sastry, S.S.: Modeling subtilin production in bacillus subtilis using stochastic hybrid systems. In: Alur, R., Pappas, G.J. (eds.) Hybrid Systems: Computation and Control. LNCS, vol. 2993, pp. 417–431. Springer, Berlin (2004)
53. Jacod, J., Skorokhod, A.V.: Jumping Markov processes. Ann. Inst. H. Poincaré Probab. Statist. **32**, 11–67 (1996)
54. Jamison, B., Orey, S.: Markov chains recurrent in the sense of Harris Z. Wahrsch. Verw. Gebiete **8**, 41–48 (1967)
55. Kallenberg, O.: Foundations of Modern Probability, 2nd edn. Springer, New York (2002)
56. Kato, T.: On the semi-groups generated by Kolmogoroff's differential equations. J. Math. Soc. Jpn **6**, 1–15 (1954)
57. Kokko, H., Jennions, M.D., Brooks, R.: Unifying and testing models of sexual selection. Annu. Rev. Ecol. Evol. Syst. **37**, 43–66 (2006)
58. Komorowski T., Tyrcha, J.: Asymptotic properties of some Markov operators. Bull. Pol. Ac.: Math. **37**, 221–228 (1989)
59. Kouretas, P., Koutroumpas, K., Lygeros, J., Lygerou, Z.: Stochastic hybrid modeling of biochemical processes. In: Cassandras, C.G., Lygeros, J. (eds.) Stochastic Hybrid Systems. Control Engineering Series, vol. 24, pp. 221–248. CRC Press, Boca Raton, FL (2007)
60. Kuno, E.: Simple mathematical models to describe the rate of mating in insect populations. Res. Popul. Ecol. **20**, 50–60 (1978)
61. Lasota, A., Mackey, M.C.: Globally asymptotic properties of proliferating cell populations. J. Math. Biol. **19**, 43–62 (1984)
62. Lasota, A., Mackey, M.C.: Chaos, Fractals, and Noise, Applied Mathematical Sciences, vol. 97. Springer, New York (1994)
63. Lasota, A., Yorke, J.A.: Exact dynamical systems and the Frobenius-Perron operator. Trans. AMS **273**, 375–384 (1982)
64. Lebowitz, J.L., Rubinow, S.L.: A theory for the age and generation time distribution of microbial population. J. Math. Biol. **1**, 17–36 (1974)
65. Lipniacki, T., Paszek, P., Marciniak-Czochra, A., Brasier, A.R., Kimmel, M.: Transcriptional stochasticity in gene expression. J. Theoret. Biol. **238**, 348–367 (2006)
66. Lotz, H.P.: Uniform convergence of operators on L^∞ and similar spaces. Math Z **190**, 207–220 (1985)
67. Lumer, G., Phillips, R.S.: Dissipative operators in a Banach space. Pacific J. Math. **11**, 679–698 (1961)
68. Mackey, M.C., Rudnicki, R.: Global stability in a delayed partial differential equation describing cellular replication. J. Math. Biol. **33**, 89–109 (1994)

69. Mackey, M.C., Santillán, M., Tyran-Kamińska, M., Zeron, E.S.: Simple Mathematical Models of Gene Regulatory Dynamics. Lecture Notes on Mathematical Modelling in the Life Sciences. Springer, Cham (2016)
70. Mackey, M.C., Tyran-Kamińska, M.: Dynamics and density evolution in piecewise deterministic growth processes. Ann. Polon. Math. **94**, 111–129 (2008)
71. Mackey, M.C., Tyran-Kamińska, M., Yvinec, R.: Molecular distributions in gene regulatory dynamics. J. Theoret. Biol. **274**, 84–96 (2011)
72. Mackey, M.C., Tyran-Kamińska, M., Yvinec, R.: Dynamic behavior of stochastic gene expression models in the presence of bursting. SIAM J. Appl. Math. **73**, 1830–1852 (2013)
73. Mackey, M.C., Tyran-Kamińska, M.: The limiting dynamics of a bistable molecular switch with and without noise. J. Math. Biol. **73**, 367–395 (2016)
74. McKendrick, A.G.: Application of mathematics to medical problems. Proc. Edinb. Math. Soc. **14**, 98–130 (1926)
75. Méndez, V., Campos, D., Pagonabarraga, I., Fedotov, S.: Density-dependent dispersal and population aggregation patterns. J. Theoret. Biol. **309**, 113–120 (2012)
76. Metz, J.A.J., Diekmann, O. (eds.): The Dynamics of Physiologically Structured Populations. Springer Lecture Notes in Biomathematics, vol. 68. Springer, New York (1986)
77. Miękisz, J., Szymańska, P.: Gene expression in self-repressing system with multiple gene copies. Bull. Math. Biol. **75**, 317–330 (2013)
78. Mokhtar-Kharroubi, M., Rudnicki, R.: On asymptotic stability and sweeping of collisionless kinetic equations. Acta Appl. Math. **147**, 19–38 (2017)
79. Nummelin, E.: General Irreducible Markov Chains and Non-negative Operators, Cambridge Tracts in Mathematics, vol. 83. Cambridge University Press, Cambridge (1984)
80. Ochab-Marcinek, A., Tabaka, M.: Transcriptional leakage versus noise: a simple mechanism of conversion between binary and graded response in autoregulated genes. Phys. Rev. E **91**, 012704 (2015)
81. Othmer, H.G., Dunbar, S.R., Alt, W.: Models of dispersal in biological systems. J. Math. Biol. **26**, 263–298 (1988)
82. Othmer, H.G., Hillen, T.: The diffusion limit of transport equations II: Chemotaxis equations. SIAM J. Appl. Math. **62**, 1222–1250 (2002)
83. Phillips, R.S.: Semi-groups of positive contraction operators. Czechoslovak Math. J. **12**, 294–313 (1962)
84. Pichór, K.: Asymptotic stability of a partial differential equation with an integral perturbation. Ann. Polon. Math. **68**, 83–96 (1998)
85. Pichór, K.: Asymptotic stability and sweeping of substochastic semigroups. Ann. Polon. Math. **103**, 123–134 (2012)
86. Pichór, K., Rudnicki, R.: Stability of Markov semigroups and applications to parabolic systems. J. Math. Anal. Appl. **215**, 56–74 (1997)
87. Pichór, K., Rudnicki, R.: Continuous Markov semigroups and stability of transport equations. J. Math. Anal. Appl. **249**, 668–685 (2000)
88. Pichór, K., Rudnicki, R.: Asymptotic behaviour of Markov semigroups and applications to transport equations. Bull. Pol. Ac.: Math. **45**, 379–397 (1997)
89. Pichór, K., Rudnicki, R.: Asymptotic decomposition of substochastic operators and semigroups. J. Math. Anal. Appl. **436**, 305–321 (2016)
90. Pichór, K., Rudnicki, R.: Asymptotic decomposition of substochastic semigroups and applications. Stochastics and Dynamics, (accepted)
91. Pichór, K., Rudnicki, R.: Stability of stochastic semigroups and applications to Stein's neuronal model. Discrete Contin. Dyn. Syst. B, (accepted)
92. Pichór, K., Rudnicki, R., Tyran-Kamińska, M.: Stochastic semigroups and their applications to biological models. Demonstratio Mathematica **45**, 463–495 (2012)
93. Rotenberg, M.: Transport theory for growing cell populations. J. Theoret. Biol. **103**, 181–199 (1983)
94. Rubinow, S.I.: A maturity time representation for cell populations. Biophys. J. **8**, 1055–1073 (1968)

95. Rudnicki, R.: On asymptotic stability and sweeping for Markov operators. Bull. Pol. Ac.: Math. **43**, 245–262 (1995)
96. Rudnicki, R.: Stochastic operators and semigroups and their applications in physics and biology. In: Banasiak, J., Mokhtar-Kharroubi, M. (eds.) Evolutionary Equations with Applications in Natural Sciences. Lecture Notes in Mathematics, vol. 2126, pp. 255–318. Springer, Heidelberg (2015)
97. Rudnicki, R., Pichór, K.: Markov semigroups and stability of the cell maturation distribution. J. Biol. Syst. **8**, 69–94 (2000)
98. Rudnicki, R., Tiuryn, J., Wójtowicz, D.: A model for the evolution of paralog families in genomes. J. Math. Biol. **53**, 759–770 (2006)
99. Rudnicki, R., Tiuryn, J.: Size distribution of gene families in a genome. Math. Models Methods Appl. Sci. **24**, 697–717 (2014)
100. Rudnicki, R., Tomski, A.: On a stochastic gene expression with pre-mRNA, mRNA and protein contribution. J. Theoret. Biol. **387**, 54–67 (2015)
101. Rudnicki, R., Tyran-Kamińska, M.: Piecewise deterministic Markov processes in biological models. Semigroups of Operators Theory And Applications. Springer Proceedings in Mathematics & Statistics 113, pp. 235–255. Springer, Cham (2015)
102. Rudnicki, R., Wieczorek, R.: Mathematical models of phytoplankton dynamics. In: Russo R. (ed.) Aquaculture I. Dynamic Biochemistry, Process Biotechnology and Molecular Biology, vol. 2 (Special Issue 1), pp. 55–63 (2008)
103. Rudnicki, R., Wieczorek, R.: Phytoplankton dynamics: from the behaviour of cells to a transport equation. Math. Model. Nat. Phenom. **1**, 83–100 (2006)
104. Rudnicki, R., Zwoleński, P.: Model of phenotypic evolution in hermaphroditic populations. J. Math. Biol. **70**, 1295–1321 (2015)
105. Sharpe, F.R., Lotka, A.J.: A problem in age-distributions. Philos. Mag. **21**, 435–438 (1911)
106. Singh, A., Hespanha, J.P.: Stochastic hybrid systems for studying biochemical processes. Philos. Trans. R. Soc. A **368**, 4995–5011 (2010)
107. Stein, R.B.: A theoretical analysis of neuronal variability. Biophys. J. **5**, 173–194 (1965)
108. Stein, R.B.: Some models of neuronal variability. Biophys. J. **7**, 37–68 (1967)
109. Stroock, D.W.: Some stochastic processes which arise from a model of the motion of a bacterium. Z. Wahrscheinlichkeitstheorie verw. Gebiete **28**, 305–315 (1974)
110. Tuckwell, H.C.: Stochastic Processes in the Neurosciences. SIAM, Philadelphia (1989)
111. Tuckwell, H.C.: Synaptic transmission in a model for stochastic neural activity. J. Theoret. Biol. **77**, 65–81 (1979)
112. Tyson, J.J., Hannsgen, K.B.: Cell growth and division: a deterministic/probabilistic model of the cell cycle. J. Math. Biol. **23**, 231–246 (1986)
113. Tyran-Kamińska, M.: Substochastic semigroups and densities of piecewise deterministic Markov processes. J. Math. Anal. Appl. **357**, 385–402 (2009)
114. Tyran-Kamińska, M.: Ergodic theorems and perturbations of contraction semigroups. Studia Math. **195**, 147–155 (2009)
115. Tyrcha, J.: Asymptotic stability in a generalized probabilistic/deterministic model of the cell cycle. J. Math. Biol. **26**, 465–475 (1988)
116. Voigt, J.: On substochastic C_0-semigroups and their generators. Transp. Theory Statist. Phys. **16**, 453–466 (1987)
117. Voigt, J.: On resolvent positive operators and positive C_0-semigroups on AL-spaces. Semigroup Forum **38**, 263–266 (1989)
118. Webb, G.W.: Theory of Nonlinear Age-Dependent Population Dynamics. Marcel Dekker, New York (1985)
119. Yin, G.G., Zhu, C.: Hybrid Switching Diffusions: Properties and Applications. Stochastic Modelling and Applied Probability, vol. 63. Springer, New York (2010)
120. Yosida, K.: On the differentiability and the representation of one-parameter semi-group of linear operators. J. Math. Soc. Jpn **1**, 15–21 (1948)
121. Yu, J., Xiao, J., Ren, X., Lao, K., Xie, X.: Probing gene expression in live cells, one protein molecule at a time. Science **311**, 1600–1603 (2006)

122. Zeiser, S., Franz, U., Müller, J., Liebscher, V.: Hybrid modeling of noise reduction by a negatively autoregulated system. Bull. Math. Biol. **71**, 1006–1024 (2009)
123. Zeiser, S., Franz, U., Liebscher, V.: Autocatalytic genetic networks modeled by piecewise-deterministic Markov processes. J. Math. Biol. **60**, 207–246 (2010)

Index

© The Author(s) 2017
R. Rudnicki and M. Tyran-Kamińska, *Piecewise Deterministic Processes
in Biological Models*, SpringerBriefs in Mathematical Methods,
DOI 10.1007/978-3-319-61295-9

Printed in the United States
By Bookmasters